IN ASSOCIATION WITH

SQA

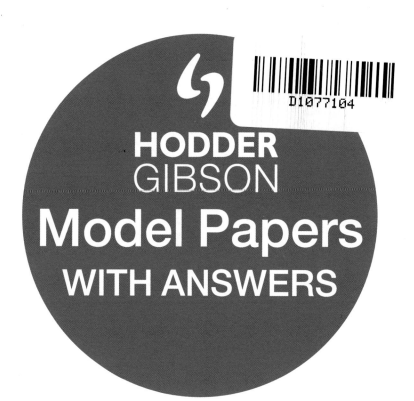

HODDER GIBSON

Model Papers
WITH ANSWERS

PLUS: Official SQA Specimen Paper & 2015 Past Paper With Answers

Higher for CfE
Biology

2014 Specimen Question Paper, Model Papers & 2015 Exam

HODDER GIBSON
AN HACHETTE UK COMPANY

This book contains the official 2014 SQA Specimen Question Paper and 2015 Exam for Higher for CfE Biology, with associated SQA approved answers modified from the official marking instructions that accompany the paper.

In addition the book contains model papers, together with answers, plus study skills advice. These papers, some of which may include a limited number of previously published SQA questions, have been specially commissioned by Hodder Gibson, and have been written by experienced senior teachers and examiners in line with the new Higher for CfE syllabus and assessment outlines, Spring 2014. This is not SQA material but has been devised to provide further practice for Higher for CfE examinations in 2015 and beyond.

Hodder Gibson is grateful to the copyright holders, as credited on the final page of the Answer Section, for permission to use their material. Every effort has been made to trace the copyright holders and to obtain their permission for the use of copyright material. Hodder Gibson will be happy to receive information allowing us to rectify any error or omission in future editions.

Hachette UK's policy is to use papers that are natural, renewable and recyclable products and made from wood grown in sustainable forests. The logging and manufacturing processes are expected to conform to the environmental regulations of the country of origin.

Orders: please contact Bookpoint Ltd, 130 Park Drive, Milton Park, Abingdon, Oxon OX14 4SE. Telephone: (44) 01235 827720. Fax: (44) 01235 400454. Lines are open 9.00–5.00, Monday to Saturday, with a 24-hour message answering service. Visit our website at www.hoddereducation.co.uk. Hodder Gibson can be contacted direct on: Tel: 0141 848 1609; Fax: 0141 889 6315; email: hoddergibson@hodder.co.uk

This collection first published in 2015 by
Hodder Gibson, an imprint of Hodder Education,
An Hachette UK Company
2a Christie Street
Paisley PA1 1NB

Typeset by Aptara, Inc.

Printed in the UK

A catalogue record for this title is available from the British Library

ISBN: 978-1-4718-6068-3

3 2 1

2016 2015

Introduction

Study Skills – what you need to know to pass exams!

Pause for thought

Many students might skip quickly through a page like this. After all, we all know how to revise. Do you really though?

Think about this:

"IF YOU ALWAYS DO WHAT YOU ALWAYS DO, YOU WILL ALWAYS GET WHAT YOU HAVE ALWAYS GOT."

Do you like the grades you get? Do you want to do better? If you get full marks in your assessment, then that's great! Change nothing! This section is just to help you get that little bit better than you already are.

There are two main parts to the advice on offer here. The first part highlights fairly obvious things but which are also very important. The second part makes suggestions about revision that you might not have thought about but which WILL help you.

Part 1

DOH! It's so obvious but …

Start revising in good time

Don't leave it until the last minute – this will make you panic.

Make a revision timetable that sets out work time AND play time.

Sleep and eat!

Obvious really, and very helpful. Avoid arguments or stressful things too – even games that wind you up. You need to be fit, awake and focused!

Know your place!

Make sure you know exactly **WHEN and WHERE** your exams are.

Know your enemy!

Make sure you know what to expect in the exam.

How is the paper structured?

How much time is there for each question?

What types of question are involved?

Which topics seem to come up time and time again?

Which topics are your strongest and which are your weakest?

Are all topics compulsory or are there choices?

Learn by DOING!

There is no substitute for past papers and practice papers – they are simply essential! Tackling this collection of papers and answers is exactly the right thing to be doing as your exams approach.

Part 2

People learn in different ways. Some like low light, some bright. Some like early morning, some like evening or night. Some prefer warm, some prefer cold. But everyone uses their BRAIN and the brain works when it is active. Passive learning – sitting gazing at notes – is the most INEFFICIENT way to learn anything. Below you will find tips and ideas for making your revision more effective and maybe even more enjoyable. What follows gets your brain active, and active learning works!

Activity 1 – Stop and review

Step 1

When you have done no more than 5 minutes of revision reading STOP!

Step 2

Write a heading in your own words which sums up the topic you have been revising.

Step 3

Write a summary of what you have revised in no more than two sentences. Don't fool yourself by saying, "I know it, but I cannot put it into words". That just means you don't know it well enough. If you cannot write your summary, revise that section again, knowing that you must write a summary at the end of it. Many of you will have notebooks full of blue/black ink writing. Many of the pages will not be especially attractive or memorable so try to liven them up a bit with colour as you are reviewing and rewriting. **This is a great memory aid, and memory is the most important thing.**

Activity 2 – Use technology!

Why should everything be written down? Have you thought about "mental" maps, diagrams, cartoons and colour to help you learn? And rather than write down notes, why not record your revision material?

What about having a text message revision session with friends? Keep in touch with them to find out how and what they are revising and share ideas and questions.

Why not make a video diary where you tell the camera what you are doing, what you think you have learned and what you still have to do? No one has to see or hear it, but the process of having to organise your thoughts in a formal way to explain something is a very important learning practice.

Be sure to make use of electronic files. You could begin to summarise your class notes. Your typing might be slow, but it will get faster and the typed notes will be easier to read than the scribbles in your class notes. Try to add different fonts and colours to make your work stand out. You can easily Google relevant pictures, cartoons and diagrams which you can copy and paste to make your work more attractive and **MEMORABLE**.

Activity 3 – This is it. Do this and you will know lots!

Step 1

In this task you must be very honest with yourself! Find the SQA syllabus for your subject (www.sqa.org.uk). Look at how it is broken down into main topics called MANDATORY knowledge. That means stuff you MUST know.

Step 2

BEFORE you do ANY revision on this topic, write a list of everything that you already know about the subject. It might be quite a long list but you only need to write it once. It shows you all the information that is already in your long-term memory so you know what parts you do not need to revise!

Step 3

Pick a chapter or section from your book or revision notes. Choose a fairly large section or a whole chapter to get the most out of this activity.

With a buddy, use Skype, Facetime, Twitter or any other communication you have, to play the game "If this is the answer, what is the question?". For example, if you are revising Geography and the answer you provide is "meander", your buddy would have to make up a question like "What is the word that describes a feature of a river where it flows slowly and bends often from side to side?".

Make up 10 "answers" based on the content of the chapter or section you are using. Give this to your buddy to solve while you solve theirs.

Step 4

Construct a wordsearch of at least 10 × 10 squares. You can make it as big as you like but keep it realistic. Work together with a group of friends. Many apps allow you to make wordsearch puzzles online. The words and phrases can go in any direction and phrases can be split. Your puzzle must only contain facts linked to the topic you are revising. Your task is to find 10 bits of information to hide in your puzzle, but you must not repeat information that you used in Step 3. DO NOT show where the words are. Fill up empty squares with random letters. Remember to keep a note of where your answers are hidden but do not show your friends. When you have a complete puzzle, exchange it with a friend to solve each other's puzzle.

Step 5

Now make up 10 questions (not "answers" this time) based on the same chapter used in the previous two tasks. Again, you must find NEW information that you have not yet used. Now it's getting hard to find that new information! Again, give your questions to a friend to answer.

Step 6

As you have been doing the puzzles, your brain has been actively searching for new information. Now write a NEW LIST that contains only the new information you have discovered when doing the puzzles. Your new list is the one to look at repeatedly for short bursts over the next few days. Try to remember more and more of it without looking at it. After a few days, you should be able to add words from your second list to your first list as you increase the information in your long-term memory.

FINALLY! Be inspired...

Make a list of different revision ideas and beside each one write **THINGS I HAVE** tried, **THINGS I WILL** try and **THINGS I MIGHT** try. Don't be scared of trying something new.

And remember – "FAIL TO PREPARE AND PREPARE TO FAIL!"

Higher Biology

The practice papers in this book give an overall and comprehensive coverage of assessment of **Knowledge** and **Scientific Inquiry** for the new CfE Higher Biology.

We recommend that you download and print a copy of Higher Biology Course Assessment Specification (CAS) pages 8–18 from the SQA website at www.sqa.org.uk.

The course

The Higher Biology Course consists of three National Units. These are DNA and the Genome, Metabolism and Survival and Sustainability and Interdependence. In each of the units you will be assessed on your ability to demonstrate and apply knowledge of Biology and to demonstrate and apply skills of scientific inquiry. Candidates must also complete an Assignment in which they research a topic in biology and write it up as a report. They also take a Course examination.

How the course is graded

To achieve a course award for Higher Biology you must pass all three National Unit Assessments which will be assessed by your school or college on a pass or fail basis. The grade you get depends on the following two course assessments, which are set and graded by SQA.

1. An 800–1200 word report based on an Assignment, which is worth 17% of the grade. The Assignment is marked out of 20 marks, with 15 of the marks being for scientific inquiry skills and 5 marks for the application of knowledge.

2. A written course examination is worth the remaining 83% of the grade. The examination is marked out of 100 marks, most of which are for the demonstration and application of knowledge although there are also marks available for skills of scientific inquiry.

This book should help you practice the examination part! To pass Higher Biology with a C grade you will need about 50% of the 120 marks available for the Assignment and the Course Examination combined. For a B you will need roughly 60% and, for an A, roughly 70%.

The course examination

The Course Examination is a single question paper in two sections.

- The first section is an objective test with 20 multiple choice items for 20 marks.

- The second section is a mix of restricted and extended response questions worth between 2 and 9 marks each for a total of 80 marks. The majority of the marks test knowledge with an emphasis on the application of knowledge. The remainder, test the application of scientific inquiry, analysis and problem solving skills. There will usually be opportunity to comment on or suggest modifications to an experimental situation.

Altogether, there are 100 marks and you will have 2 hours and 30 minutes to complete the paper. The majority of the marks will be straightforward and linked to grade C but some questions are more demanding and are linked to grade A.

General tips and hints

You should have a copy of the Course Assessment Specification (CAS) for Higher Biology but, if you haven't got one, make sure to download it from the SQA website. This document tells you what can be tested in your examination. It is worth spending some time on this document. This book contains four practice Higher examination papers. One is the SQA specimen paper and there are three further model papers. Each paper has been carefully assembled to be as similar as possible to a typical Higher Biology Paper. Notice how similar they all are in the way in which they are laid out and the types of question they ask – your own course examination is going to be very similar as well, so the value of the papers is obvious! Each paper can be attempted in its entirety or groups of questions on a particular topic or skill area can be attempted. If you are trying a whole examination paper from this book, give yourself 2 hours and 30 minutes maximum to complete it. The questions in each paper are laid out in Unit order. Make sure that you spend time in using the answer section to mark your own work – it is especially useful if you can get someone to help you with this.

The marking instructions give acceptable answers with alternatives. You could even grade your work on an A–D basis. The following hints and tips are related to examination techniques as well as avoiding common mistakes. Remember that if you hit problems with a question, you should ask your teacher for help.

Section 1

20 multiple-choice items **20 marks**

- Answer on a grid.

- Do not spend more than 30 minutes on this section.

- Some individual questions might take longer to answer than others – this is quite normal and make sure you use scrap paper if a calculation or any working is needed.

- Some questions can be answered instantly– again, this is normal.

- Do not leave blanks – complete the grid for each question as you work through.

- Try to answer each question in your head without looking at the options. If your answer is there you are home and dry!

- If you are not certain, choose the answer that seemed most attractive on first reading the answer options.

- If you are guessing, try to eliminate options before making your guess. If you can eliminate three – you are left with the correct answer even if you do not recognise it!

Section 2

Restricted and extended response **80 marks**

- Spend about 2 hours on this section.

- Answer on the question paper. Try to write neatly and keep your answers on the support lines if possible – the lines are designed to take the full answer!

- A clue to answer length is the mark allocation – most questions are restricted to 1 mark and the answer can be quite short. If there are 2–4 marks available, your answer will need to be extended and may well have two, three or even four parts.

- The questions are usually laid out in Unit sequence but remember some questions are designed to cover more than one Unit.

- The C-type questions usually start with "State", "Identify", "Give" or "Name" and often need only a word or two in response. They will usually be for 1 mark each.

- Questions that begin with " Explain" and "Describe" are usually A types and are likely to have more than one part to the full answer. You will usually have to write a sentence or two and there may be 2 or even 3 marks available.

- Make sure you read questions over twice before trying to answer – there is often very important information within the question and you are unlikely to be short of time in this examination.

- Using abbreviations like DNA and ATP is fine and the bases of DNA can be given as A, T, G and C. The Higher Biology Course Assessment Specification (CAS) will give you the acceptable abbreviations.

- Don't worry that a few questions are in unfamiliar contexts, that's the idea! Just keep calm and read the questions carefully.

- If a question contains a choice, be sure to spend a minute or two making the best choice for you.

- In experimental questions, you must be aware of what variables are, why controls are needed and how reliability and validity might be improved. It is worth spending time on these ideas – they are essential and will come up year after year.

- Some candidates like to use a highlighter pen to help them focus on the essential points of longer questions – this is a great technique.

- Remember that a conclusion can be seen from data, whereas an explanation will usually require you to supply some background knowledge as well.

- Remember to "use values from the graph" when describing graphical information in words if you are asked to do so.

- Plot graphs carefully and join the plot points using a ruler. Include zeros on your scale where appropriate and use the data table headings for the axes labels.

- Look out for graphs with two Y-axes – these need extra special concentration and anyone can make a mistake!

- If there is a space for calculation given – you will very likely need to use it! A calculator is essential.

- The main types of calculation tend to be ratios, averages, percentages and percentage change – make sure you can do these common calculations.

- Answers to calculations will not usually have more than two decimal places.

- Do not leave blanks. Always have a go, using the language in the question if you can.

Good luck!

Remember that the rewards for passing Higher Biology are well worth it! Your pass will help you get the future you want for yourself. In the exam, be confident in your own ability. If you're not sure how to answer a question, trust your instincts and just give it a go anyway.

Keep calm and don't panic! GOOD LUCK!

2014 Specimen Question Paper

National Qualifications
SPECIMEN ONLY

SQ04/H/02

Biology
Section 1—Questions

Date — Not applicable

Duration — 2 hours and 30 minutes

Instructions for the completion of Section 1 are given on *Page two* of your question and answer booklet.

Record your answers on the answer grid on *Page three* of your question and answer booklet.

Before leaving the examination room you must give your question and answer booklet to the Invigilator; if you do not, you may lose all the marks for this paper.

SECTION 1 — 20 marks
Attempt ALL questions

1. The genetic material in human mitochondria is arranged as

    A   linear chromosomes

    B   circular plasmids

    C   circular chromosomes

    D   inner membranes.

2. The main components of a ribosome are

    A   mRNA and tRNA

    B   rRNA and protein

    C   mRNA and protein

    D   rRNA and mRNA.

3. The diagram below represents part of a protein molecule.

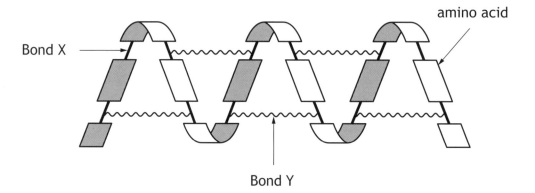

    Which line in the table below identifies bonds X and Y?

	Bond X	Bond Y
A	hydrogen	peptide
B	hydrogen	hydrogen
C	peptide	hydrogen
D	peptide	peptide

4. Types of single gene mutation are given in the list below.

    1 substitution

    2 insertion

    3 deletion

    Which of these would affect only one amino acid in the polypeptide produced?

    A   1 only

    B   2 only

    C   3 only

    D   2 and 3 only

5. Which line in the table below describes meristems?

	Found in	Type of cell present
A	animal	specialised
B	animal	unspecialised
C	plant	specialised
D	plant	unspecialised

6. The table below provides information about ancestral and modern Brassica species. The modern species have been produced by hybridisation of two ancestral species followed by a doubling of the chromosome number in the hybrids.

Brassica species	Ancestral or modern species	Crop	Diploid chromosome number (2 n)
B. oleracea	ancestral	cabbage	18
B. nigra	ancestral	black mustard	16
B. rapa	ancestral	turnip	20
B. juncea	modern	Indian Mustard	36
B. carinata	modern	Ethiopian Mustard	34
B. napus	modern	oilseed rape	38

Which of the following shows the ancestral hybridisation and the modern species produced?

A   Cabbage × turnip ⟶ oilseed rape

B   Turnip × black mustard ⟶ Ethiopian mustard

C   Turnip × cabbage ⟶ Indian mustard

D   Cabbage × black mustard ⟶ Indian mustard

7. The diagram below shows how a molecule might be biosynthesised from building blocks in a metabolic pathway.

building blocks   biosynthesised molecule

Which line in the table below describes the metabolic process shown in the diagram and the energy relationship involved in the reaction?

	Metabolic process	Energy relationship
A	anabolic	energy used
B	anabolic	energy released
C	catabolic	energy used
D	catabolic	energy released

8. The graph below shows changes in the α-amylase concentration and the starch content of a barley grain during early growth and development.

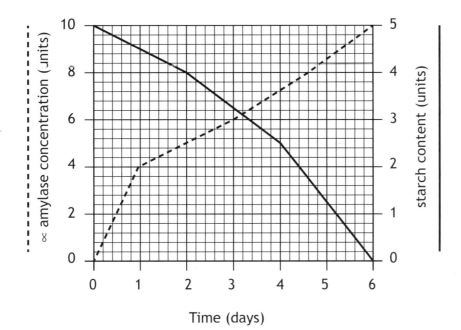

Time (days)

Identify the time by which the starch content of the barley grains had decreased by 50%.

A 2·0 days

B 3·2 days

C 4·0 days

D 6·0 days

9. The graph below shows the effect of different concentrations of a disinfectant on the number of viable bacteria in liquid culture.

Key: % concentration of disinfectant

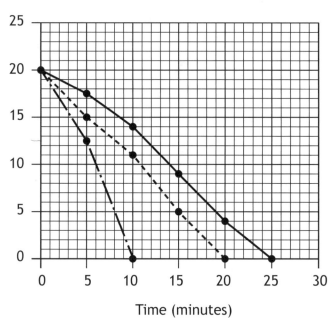

What percentage of bacteria was killed by 20% disinfectant after 5 minutes?

A   25

B   37·5

C   62·5

D   75

10. The diagram below shows a bacterial cell that has been magnified 800 times.

3·2 mm

Calculate the length of the cell in micrometres (μm).

A   0·004

B   0·04

C   0·4

D   4·0

11. The cell membrane contains pumps that actively transport substances.

Which of the following forms the major component of membrane pumps?

A   Protein

B   Phospholipid

C   Nucleic acid

D   Carbohydrate

12. Maximum oxygen uptake per kg body mass can be used as a measure of fitness. Four athletes were weighed then given a fitness test during which their maximum oxygen uptake was measured.

Which line in the table below shows results for the least fit athlete?

Athlete	Body mass (kg)	Maximum oxygen uptake (litres per minute)
A	60	3·6
B	55	3·6
C	60	3·7
D	55	3·7

13. The list below gives some adaptations of weed plants.

1   high seed output

2   possession of storage organs

3   vegetative reproduction

4   long term seed viability

Which of these are competitive adaptations of annual weeds?

A   1 and 2 only

B   1 and 4 only

C   2 and 3 only

D   2 and 4 only

14. The table below gives measurements relating to productivity in a field of wheat grown to produce grain for making bread.

Measurement	Productivity (kg dry mass per hectare per year)
plant biomass	11 250
grain yield	4500

What is the harvest index of this wheat crop?

A   0·4

B   2·5

C   6750

D   15750

15. The action spectrum of photosynthesis is a measure of the ability of plants to

A   absorb all wavelengths of light

B   absorb light of different intensities

C   use light to build up food

D   use light of different wavelengths for photosynthesis.

16. The flow chart below shows the energy flow in a field of potatoes during one year.

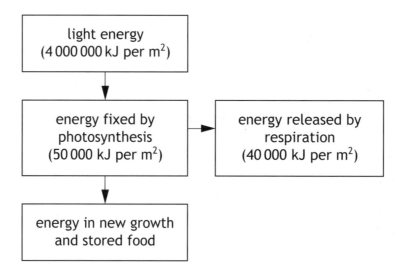

What is the percentage of the available light energy present in new growth and stored food in the potato crop?

A   2·25

B   1·25

C   0·25

D   1·00

17. The diagram below represents part of the Calvin cycle within a chloroplast.

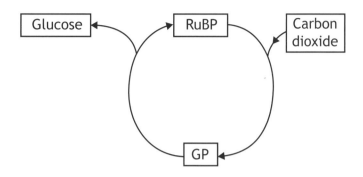

Which line in the table below shows the effect of decreasing CO_2 availability on the concentrations of RuBP and GP in the cycle?

	RuBP concentration	GP concentration
A	decrease	decrease
B	increase	increase
C	decrease	increase
D	increase	decrease

18. The list below describes observed behaviour of pigs on a farm.

1   Stereotypic flicking of the head

2   Repeated wounding of other pigs by biting

3   Lying in a position which does not allow suckling

Which of these behaviours indicate poor animal welfare?

A   1 and 2 only

B   1 and 3 only

C   2 and 3 only

D   1, 2 and 3

19. Adult beef tapeworms live in the intestine of humans. Segments of the adult worm are released in the faeces. Embryos that develop from them remain viable for five months. The embryos may be eaten by cattle and develop in their muscle tissue.

Which row in the table below identifies the roles of the human, tapeworm embryo and cattle?

		Role	
	human	*tapeworm embryo*	*cattle*
A	host	resistant stage	secondary host
B	host	vector	secondary host
C	secondary host	vector	host
D	secondary host	resistant stage	vector

20. Ostriches are large birds that live on open plains in Africa. They divide their time between feeding on vegetation and raising their heads to look for predators.

The graphs below show the results of a study on the effect of group size in ostriches on their behaviour.

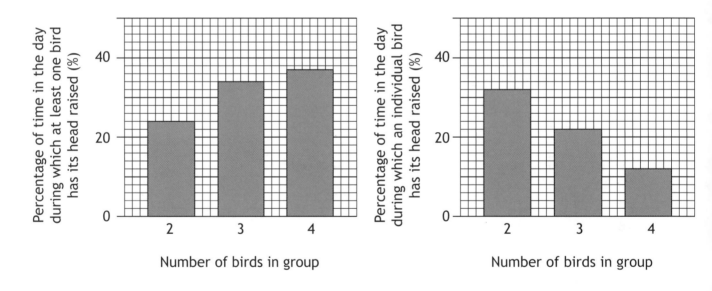

Which of the following is a valid conclusion from these results?

In larger groups, an individual ostrich spends

A   less time with its head raised so the group is less likely to see predators

B   less time with its head raised but the group is more likely to see predators

C   more time with its head raised so the group is more likely to see predators

D   more time with its head raised but the group is less likely to see predators.

[END OF SECTION 1. NOW ATTEMPT THE QUESTIONS IN SECTION 2 OF YOUR QUESTION AND ANSWER BOOKLET]

H

National
Qualifications
SPECIMEN ONLY

Mark

SQ04/H/01

Biology
Section 1 — Answer Grid and Section 2

Date — Not applicable

Duration — 2 hours and 30 minutes

Fill in these boxes and read what is printed below.

Full name of centre

Town

Forename(s)

Surname

Number of seat

Date of birth

Day	Month	Year
D D	M M	Y Y

Scottish candidate number

Total marks — 100

SECTION 1 — 20 marks

Attempt ALL questions.

Instructions for completion of Section 1 are given on *Page two*.

SECTION 2 — 80 marks

Attempt ALL questions.

Write your answers clearly in the spaces provided in this booklet. Additional space for answers and rough work is provided at the end of this booklet. If you use this space you must clearly identify the question number you are attempting. Any rough work must be written in this booklet. You should score through your rough work when you have written your final copy.

Use **blue** or **black** ink.

Before leaving the examination room you must give this booklet to the Invigilator; if you do not you may lose all the marks for this paper.

SECTION 1— 20 marks

The questions for Section 1 are contained in the question paper SQ04/H/02.
Read these and record your answers on the answer grid on Page three opposite.
Do NOT use gel pens.

1.   The answer to each question is **either** A, B, C or D. Decide what your answer is, then fill in the appropriate bubble (see sample question below).

2.   There is **only one correct** answer to each question.

3.   Any rough working should be done on the additional space for answers and rough work at the end of this booklet.

Sample Question

The thigh bone is called the

    A   humerus

    B   femur

    C   tibia

    D   fibula.

The correct answer is **B**—femur. The answer **B** bubble has been clearly filled in (see below).

Changing an answer

If you decide to change your answer, cancel your first answer by putting a cross through it (see below) and fill in the answer you want. The answer below has been changed to **D**.

If you then decide to change back to an answer you have already scored out, put a tick (✓) to the **right** of the answer you want, as shown below:

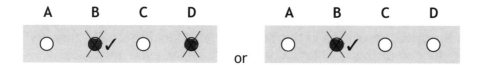

or

SECTION 1 — Answer Grid

	A	B	C	D
1	○	○	○	○
2	○	○	○	○
3	○	○	○	○
4	○	○	○	○
5	○	○	○	○
6	○	○	○	○
7	○	○	○	○
8	○	○	○	○
9	○	○	○	○
10	○	○	○	○
11	○	○	○	○
12	○	○	○	○
13	○	○	○	○
14	○	○	○	○
15	○	○	○	○
16	○	○	○	○
17	○	○	○	○
18	○	○	○	○
19	○	○	○	○
20	○	○	○	○

MARKS | DO NOT WRITE IN THIS MARGIN

SECTION 2 — 80 marks

Attempt ALL questions

It should be noted that questions 8 and 14 contain a choice.

1. The diagram below shows stages in the production of three different proteins that are coded for by one gene.

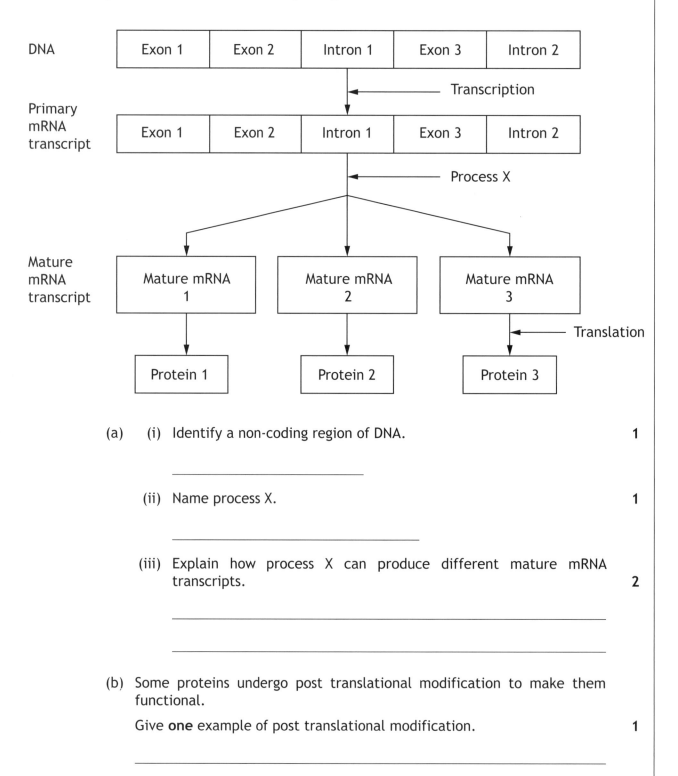

(a) (i) Identify a non-coding region of DNA.

1

_____

(ii) Name process X.

1

_____

(iii) Explain how process X can produce different mature mRNA transcripts.

2

_____

_____

(b) Some proteins undergo post translational modification to make them functional.

Give **one** example of post translational modification.

1

_____

MARKS

2. A chromosome mutation in humans can result in the formation of the Philadelphia chromosome, which is associated with a form of leukaemia.

The stages leading to the formation of a Philadelphia chromosome are shown in the diagram below.

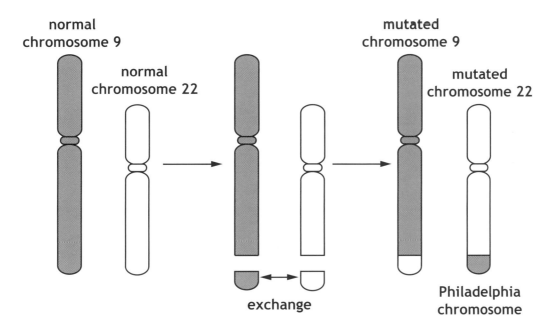

(a) Name the type of chromosome mutation, shown in the diagram, which results in the formation of a Philadelphia chromosome.

1

_____

(b) (i) The presence of a Philadelphia chromosome causes a form of leukaemia through the over-production of an enzyme.

A drug has been used to successfully treat this form of leukaemia by blocking the active site of the enzyme.

Name the type of enzyme inhibition shown by this drug.

1

_____

MARKS | DO NOT WRITE IN THIS MARGIN

2. (b) (continued)

(ii) White blood cell counts in humans normally range from 5000 to 10 000 cells per μl of blood.

The table below shows the white blood cell counts from a patient with leukaemia before and after treatment with this drug.

	Number of white blood cells (per μl blood)
Before treatment	150 000
After treatment	7500

Calculate the percentage decrease in the number of white blood cells after treatment with this drug.

Space for calculation

1

_____ %

(iii) Explain how the results suggest that the type of leukaemia in this patient was a result of the presence of a Philadelphia chromosome.

2

_____

_____

_____

MARKS | DO NOT WRITE IN THIS MARGIN

3. The polymerase chain reaction (PCR) amplifies specific sequences of DNA.

   The flow chart below shows how a sample of DNA was treated during a cycle of the PCR procedure.

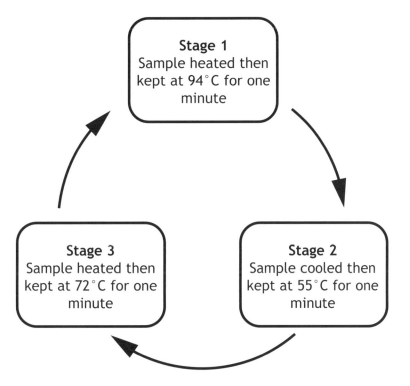

Stage 1
Sample heated then kept at 94°C for one minute

Stage 2
Sample cooled then kept at 55°C for one minute

Stage 3
Sample heated then kept at 72°C for one minute

(a) Explain the purpose of the different heat treatments in Stage 1 and Stage 2.

2

_____

_____

_____

MARKS | DO NOT WRITE IN THIS MARGIN

3. (continued)

(b) The number of DNA molecules doubles during each cycle of the PCR procedure.

Calculate the number of cycles needed to produce 128 copies of a single DNA molecule.

1

Space for calculation

_____ cycles

(c) The diagram below shows the contents of a tube used in PCR.

Contents of tube
— DNA
— DNA nucleotides
— primers
— enzyme and buffer

Describe the contents of a suitable control tube designed to show that primers are needed in the reaction.

1

_____

_____

(d) State **one** practical application of PCR.

1

_____

MARKS | DO NOT WRITE IN THIS MARGIN

4. The phylogenetic tree below shows the evolutionary relationship between the three domains of life into which all present day living things can be divided.

Evolutionary relationship   Domains of life

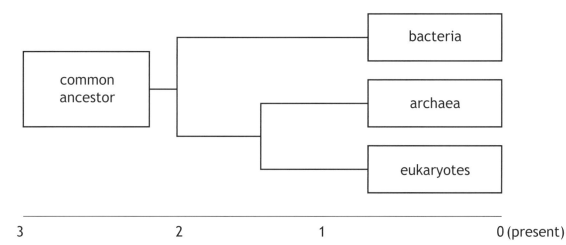

3   2   1   0 (present)

Time (billions of years ago)

(a) Name the type of data that can be used to confirm the evolutionary relationships between the domains of life shown on the diagram.

1

_____

(b) Around one billion years ago genes were transferred between archaea and bacteria.

Give the term that describes this form of gene transfer.

1

_____

MARKS | DO NOT WRITE IN THIS MARGIN

4. **(continued)**

(c) The phylogenetic tree below illustrates the evolutionary relationships between primate groups.

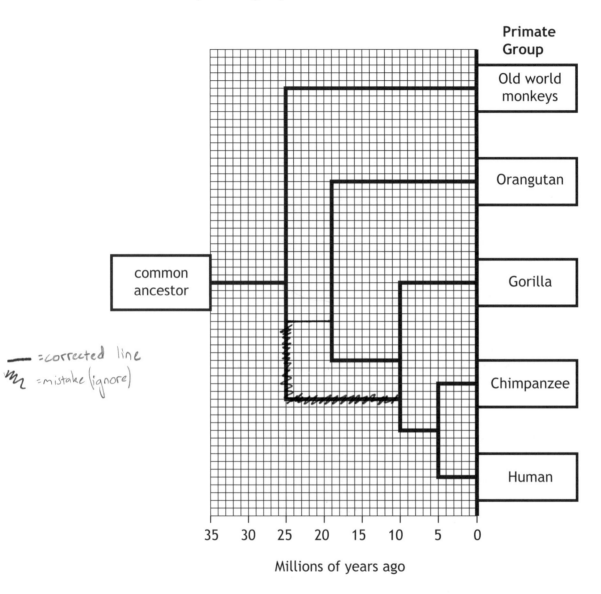

— = corrected line

m = mistake (ignore)

Millions of years ago

(i) State how long ago the last common ancestor of gorillas and old world monkeys existed.

1

_____ million years ago

MARKS | DO NOT WRITE IN THIS MARGIN

4. (c) (continued)

(ii) Humans are more closely related to chimpanzees than to orangutans.

Explain how this is known, using information from the phylogenetic tree above.

2

_____

_____

_____

MARKS DO NOT WRITE IN THIS MARGIN

5. The diagram below shows some stages in the aerobic respiration of glucose.

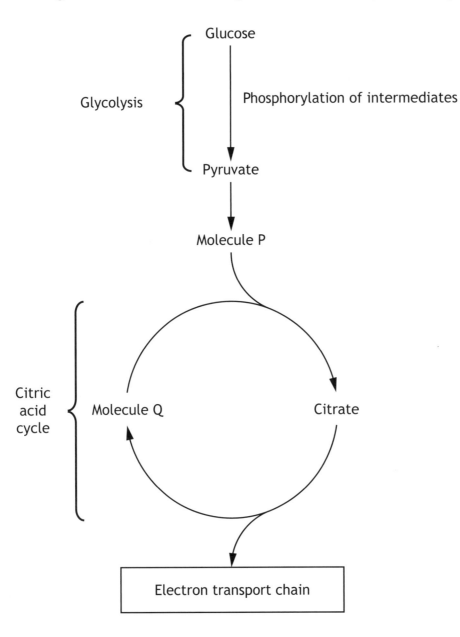

(a) Name molecules P and Q.   2

   Molecule P _____

   Molecule Q _____

(b) Explain why the phosphorylation of intermediates in glycolysis is
    referred to as an energy investment stage.   2

    _____

    _____

MARKS | DO NOT WRITE IN THIS MARGIN

5. **(continued)**

(c) Describe the role of the coenzymes NAD and FAD.   2

_____

_____

(d) People who suffer from chronic fatigue syndrome have mitochondria in which some of the proteins embedded in the inner mitochondrial membrane are damaged.

Explain how this might result in the tiredness that is a feature of this condition.   2

_____

_____

_____

6. The graph below shows the number of reported cases of hospital acquired infections (HAI) in one hospital between 2002 and 2008. The overall number of patients remained constant during this time.

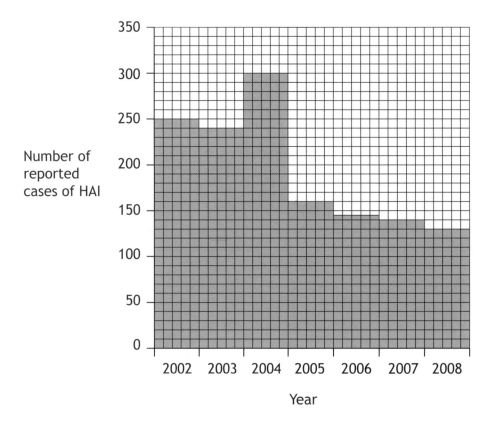

Number of reported cases of HAI

Year

(a) Using information from the graph, calculate the average decrease per year in reported cases of HAI between 2002 and 2008.

1

Space for calculation

_____ cases per year

MARKS | DO NOT WRITE IN THIS MARGIN

6. (continued)

(b) The decrease in the number of cases in 2005 was due to introduction of a new hand washing procedure at the hospital.

Predict what would happen to the number of reported cases of HAI in 2009.

Circle **one** answer and give a reason for your choice.   1

increase   decrease   stay the same

Reason _____

_____

(c) The table below shows the percentage of cases of HAI in the hospital attributed to two types of bacteria, *Clostridium* and *Staphylococcus*, between 2002 and 2008.

Bacterial types	Percentage of cases of HAI in each year attributed to bacterial types						
	2002	2003	2004	2005	2006	2007	2008
Clostridium	32	30	30	51	54	57	59
Staphylococcus	34	32	33	30	31	33	33

Using information in the table, compare the overall trend in the percentage of *Clostridium* cases with that of *Staphylococcus* cases.   2

_____

_____

(d) Using information from the graph and the table, draw a conclusion about the effectiveness of the hand washing procedure against *Staphlycoccus*. Justify your answer.   2

Conclusion _____

_____

Justification _____

_____

MARKS | DO NOT WRITE IN THIS MARGIN

6. (continued)

(e) Some bacteria form endospores to survive adverse conditions. Identify which of the two types of bacteria in the table forms endospores and give a reason for your answer.

1

Bacterial type _____

Reason _____

_____

1

MARKS | DO NOT WRITE IN THIS MARGIN

7. Mammals are regulators and can control their internal environment.

    (a) Give **one** reason why it is important for mammals to regulate their body temperature.

        1

_____

_____

    (b)   (i) Name the temperature monitoring centre in the body of a mammal.

        1

_____

      (ii) State how messages are sent from the temperature monitoring centre to the skin.

        1

_____

    (c) The blood vessels in the skin of a mammal respond to a decrease in environmental temperature.

      (i) Describe this response.

        1

_____

      (ii) Explain the effect of this response.

        1

_____

MARKS | DO NOT WRITE IN THIS MARGIN

8. Answer **either A or B**.

   A   Describe how animals survive adverse conditions.   4

   OR

   B   Describe recombinant DNA technology.   4

   Labelled diagrams may be used where appropriate.

MARKS | DO NOT WRITE IN THIS MARGIN

9. The average yield, fat and protein content of the milk from each of three breeds of dairy cattle were determined.

The results are shown in the table below.

Breed	Average milk yield per cow (kg per day)	Average fat content of milk (%)	Average protein content of milk (%)
Pure bred Holstein	44·80	4·15	3·25
F$_1$ hybrid Holstein × Normande	48·64	4·25	3·10
F$_1$ hybrid Holstein × Scandinavian Red	51·52	4·25	3·15

(a) Calculate the percentage increase in average milk yield per cow from the F$_1$ hybrid Holstein × Scandinavian Red compared to pure bred Holstein cattle.

1

Space for calculation

_____ %

(b) The fat content of milk is important for butter production.

Calculate the total fat content in the milk produced in a day from a herd of 200 F$_1$ hybrid Holstein × Normande cattle.

1

Space for calculation

_____ kg per day

MARKS | DO NOT WRITE IN THIS MARGIN

9.   (continued)

(c) Select **one** from: average milk yield per cow; average fat content of milk; or average protein content of milk.

For your choice, draw a conclusion about the effects of crossbreeding.   **1**

Choice _____

Conclusion _____

_____

(d) The development of pure breeds such as Holsteins has led to an accumulation of deleterious recessive alleles.

State the term that describes this.   **1**

_____

(e) Some F_2 offspring from crosses of F_1 hybrid Holstein × Scandinavian Red cattle will have less desirable milk-producing characteristics than their parents.

(i) Give **one** reason for this.   **1**

_____

_____

(ii) Name a process breeders would have to carry out to maintain the milk-producing characteristics of the F_1 hybrids in further generations.   **1**

_____

10. An investigation was carried out to compare the rate of photosynthesis, at different light intensities, of green algal cells immobilised into gel beads.

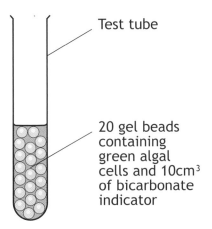

Test tube

20 gel beads containing green algal cells and 10cm³ of bicarbonate indicator

Seven tubes were set up as shown in the diagram and each positioned at a different distance from a light source to alter the light intensity.

Photosynthesis causes the bicarbonate indicator solution to change colour.

After 60 minutes, the bicarbonate indicator solution was transferred from each tube to a colorimeter.

The higher the colorimeter reading, the higher the rate of photosynthesis that has occurred in the tube.

Results are shown in the table.

Tube	Distance of tube from light source (cm)	Colorimeter reading (units)
1	25	92
2	35	92
3	50	83
4	75	32
5	100	14
6	125	6
7	200	0

MARKS | DO NOT WRITE IN THIS MARGIN

10. **(continued)**

(a) Identify the dependent variable in this investigation.   1

_____

(b) Describe how the apparatus could be improved to ensure that temperature was kept constant.   1

_____

(c) State an advantage of using algae immobilised into gel beads.   1

_____

_____

(d) Describe how the experimental procedure could be improved to increase the reliability of the results.   1

_____

_____

MARKS DO NOT WRITE IN THIS MARGIN

10. (continued)

(e) On the grid below, complete the line graph to show the colorimeter reading against distance of tube from light source.

2

(Additional graph paper if required will be found on *Page twenty-nine*)

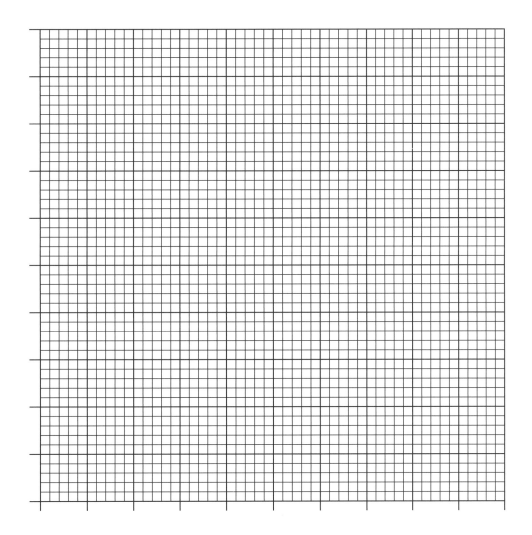

(f) From the results of this investigation, draw a conclusion about the effect of light intensity on the rate of photosynthesis.

2

_____

_____

_____

11. (a) The honey bee (*Apis mellifera*) is a social insect that lives in colonies.

The queen is the only female in a colony that reproduces. Other females are workers that collect food, maintain the colony and care for the developing offspring.

Explain the advantage to the worker bees of caring for the offspring of the queen.

2

_____

_____

(b) The graph below shows the changes in the number of honey bee hives kept by bee-keepers in the USA from 1945 to 2005.

Number of honey bee hives (millions)

Year

MARKS | DO NOT WRITE IN THIS MARGIN

11. (b) (continued)

(i) **Using values from the graph**, describe changes in the number of bee hives from 1980 to 1995.

1

_____

_____

(ii) Calculate the simplest whole number ratio of the number of bee hives in 1965 and 2005.

1

Space for calculation

_____ hives in 1965 : _____ hives in 2005

MARKS | DO NOT WRITE IN THIS MARGIN

12. The biodiversity and the genetic diversity of individual species are affected when fragments of woodland become isolated.

The diagram below illustrates habitat fragmentation of an area of woodland over time.

The shaded areas represent woodland.

time

(a)   (i)   Name **one** component of genetic diversity.   1

_____

(ii) Suggest a reason why a decrease in genetic diversity of an individual species can lead to local extinctions within habitat fragments.   1

_____

_____

(b) Suggest how habitat edge species might affect interior species as the habitat fragments become smaller.   1

_____

_____

(c) Habitat corridors can be created to remedy habitat fragmentation.

(i)   State what is meant by the term "habitat corridor".   1

_____

_____

(ii) Explain how a habitat corridor can increase biodiversity after local extinction.   1

_____

_____

MARKS DO NOT WRITE IN THIS MARGIN

13. Japanese knotweed (*Fallopia japonica*) was introduced to Britain as an ornamental plant. It grows to 3 metres in height and has large leaves. It has become naturalised and has colonised many parts of the country where it out-competes native plants.

(a) Give the term used for a naturalised species that eliminates native species.

1

_____

(b) Name **one** resource for which Japanese knotweed may outcompete the native plants.

1

_____

(c) An insect from Japan, which feeds on Japanese knotweed, has been proposed as a biological control agent.

(i) Describe **one** possible risk of introducing this insect into Britain.

1

_____

_____

_____

(ii) Describe a procedure that should be carried out to assess the risk of introducing this insect.

1

_____

_____

MARKS | DO NOT WRITE IN THIS MARGIN

14. Answer **either A or B** in the space below.

A   Describe DNA under the following headings.   9

  (i) Structure of DNA

  (ii) Replication of DNA

B   Describe the evolution of new species under the following headings.   9

  (i) Isolation and mutation

  (ii) Selection

Labelled diagrams may be used where appropriate.

[END OF SPECIMEN QUESTION PAPER]

ADDITIONAL SPACE FOR ANSWERS AND ROUGH WORK

ADDITIONAL GRAPH PAPER FOR QUESTION 10 (e)

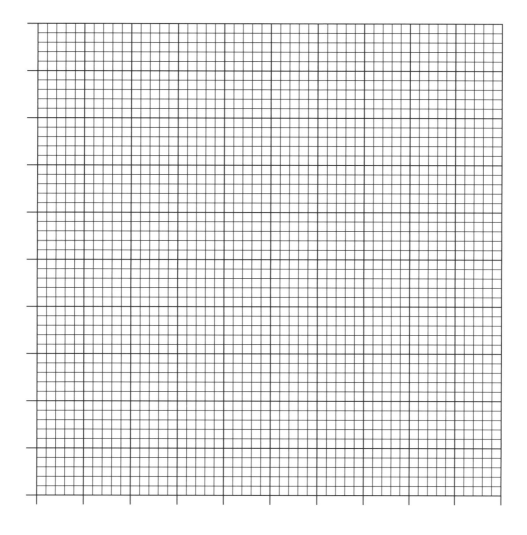

ADDITIONAL SPACE FOR ANSWERS AND ROUGH WORK

Page thirty

HIGHER FOR CfE

Model Paper 1

Whilst this Model Paper has been specially commissioned by Hodder Gibson for use as practice for the Higher (for Curriculum for Excellence) exams, the key reference documents remain the SQA Specimen Paper 2014 and SQA Past Paper 2015.

National Qualifications
MODEL PAPER 1

Biology
Section 1—Questions

Duration — 2 hours and 30 minutes

Instructions for the completion of Section 1 are given on *Page two* of your question and answer booklet.

Record your answers on the answer grid on *Page three* of your question and answer booklet.

Before leaving the examination room you must give your question and answer booklet to the Invigilator; if you do not, you may lose all the marks for this paper.

SECTION 1 — 20 marks

Attempt ALL questions

1. The diagram below shows part of a DNA molecule during replication. The bases are represented by numbers and letters.

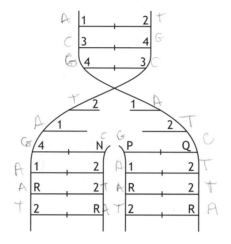

Base 1 represents adenine and base 3 represents cytosine.

Which line in the table below identifies correctly bases N, P, Q and R?

	N	P	Q	R
A	guanine	cytosine	guanine	thymine
B	cytosine	guanine	cytosine	adenine
C	guanine	cytosine	cytosine	adenine
D	cytosine	guanine	guanine	adenine

2. The diagram below shows a strand of DNA.

5′GAATTCGAT3′

Which sequence would be found on the complementary DNA strand?

A   3′GAATTCGAT5′

B   5′CTTAAGCTA3′

C   3′CTTAAGCTA5′

D   5′CUUAAGCUA3′

3. Which of the following structures would be found in both prokaryotic and eukaryotic cells?

    A   ribosome

    B   mitochondrion

    C   chloroplast

    D   nucleus

4. Which of the following best describes how phenotype is determined?

    A   by gene expression alone

    B   by gene expression influenced by intracellular factors only

    C   by gene expression influenced by extracellular factors only

    D   by gene expression influenced by both intra and extracellular factors

5. Which of the following are gene mutations?

    A   insertion, deletion and substitution

    B   substitution, duplication and translocation

    C   translocation, insertion and deletion

    D   deletion, duplication and translocation

6. The list below shows different types of ribonucleic acid (RNA).

    1   messenger RNA

    2   transfer RNA

    3   ribosomal RNA

    Which of the following types of RNA is (are) transcribed from non-coding sequences of DNA?

    A   1 only

    B   1 and 2 only

    C   2 and 3 only

    D   1, 2 and 3

7. The genetic code of a cell contained 6×10^9 base pairs of which 4% coded for proteins.

   How many DNA codons code for proteins?

   A 8×10^7

   B 8×10^8

   C 2.4×10^7

   D 2.4×10^8

8. An enzyme and the substrate it breaks down were mixed with various concentrations of copper or magnesium ions. The time taken for the complete breakdown of the substrate was measured and the results recorded in the table below.

Metal ion concentration (units)	Time taken for complete breakdown of substrate (seconds)	
	Copper ions	Magnesium ions
0	39	39
2	42	21
10	380	49
50	1480	286

Which line in the table below describes correctly the effects of high concentrations of the metal ions on the activity of this enzyme?

	High concentration of copper ions	High concentration of magnesium ions
A	promoted	promoted
B	promoted	inhibited
C	inhibited	inhibited
D	inhibited	promoted

Page four

9. Three different strains of yeast each lacked a different respiratory enzyme involved in the complete breakdown of glucose. The effect of the missing enzyme on each of the strains is shown in the table below.

Strain	Effect of missing enzyme
X	Cannot produce carbon dioxide from pyruvate
Y	Cannot synthesise pyruvate
Z	Cannot reduce oxygen to produce water

Which of the strains could produce ethanol?

A   X and Y

B   X and Z

C   Y only

D   Z only

10. Which line in the table below shows correctly the roles of proteins embedded in phospholipid membranes?

	Roles of protein		
	pumps	pores	enzymes
A	x	x	✓
B	✓	✓	x
C	✓	x	x
D	✓	✓	✓

11. In genetic engineering procedures, endonucleases are used to

A   join fragments of DNA together

B   cut DNA into specific fragments

C   seal DNA fragments into plasmids

D   add nucleotides to DNA fragments.

12. The graph below shows the effect of air temperature on the metabolic rate of two different animals.

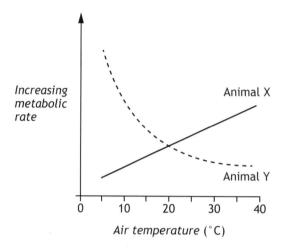

Which line in the table below identifies correctly the temperatures at which oxygen consumption will be the greatest in the tissues of each animal?

	Temperature (°C)	
	Animal X	Animal Y
A	20	20
B	40	40
C	40	5
D	5	40

13. The diagram below shows a natural interaction between two strains of fungi X and Y of different genotypes in an area Z.

Which statement describes the process occurring in area Z?

A   mutagenesis

B   selective breeding

C   transfer of plasmids

D   sexual reproduction

14. In photosynthesis, the function of carotenoid pigments is to

    A   receive light energy from chlorophyll for use in photolysis

    B   allow plants to absorb a wider range of light wavelengths

    C   increase photosynthesis at low light intensities

    D   change the capacity of chlorophyll to absorb light.

15. The energy which reaches the Earth as sunlight is 2.0×10^{12} kilojoules per hectare per year on average. The energy captured by photosynthesis is 2.0×10^{10} kilojoules per hectare per year on average.

    What percentage of available light energy is captured in photosynthesis?

    A   1%

    B   2%

    C   17%

    D   200%

16. Livestock breeders carry out backcrosses during breeding programmes.

    Backcrosses are performed to

    A   combine characteristics of separate breeds

    B   confirm the genotype of an individual

    C   maintain characteristics of a new breed

    D   increase the homozygosity of a breed.

17. Which of the following are competitive adaptations in perennial crop weeds?

    A   high seed output and storage organs

    B   storage organs and vegetative reproduction

    C   vegetative reproduction and short life cycles

    D   short life cycles and high seed output

18. The main advantage gained by using land for crops rather than livestock is that crop

    A   plants produce more food per unit area of land than livestock

    B   plants produce food with higher nutritional values than livestock

    C   production is more sustainable than livestock production

    D   plants create less environmental damage than livestock.

19. The increase in mass of a crop plant minus the loss due to respiration is called

    A   biological yield

    B   net productivity

    C   economic yield

    D   harvest index.

20. Tomato is a crop which contributes to food security in many parts of the world. The graph below shows the changes in the starch content and the mass of the pigment lycopene in ripening tomato fruits.

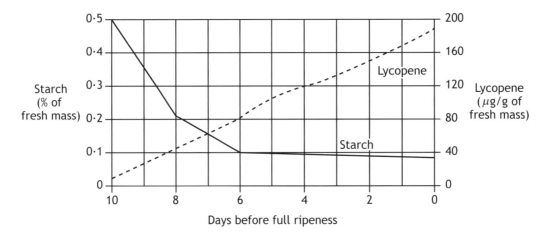

Days before full ripeness

Which of the following is a valid conclusion which can be drawn from the data in the graph?

During the ripening of tomato fruits

    A   starch is converted into lycopene

    B   the faster starch is broken down the faster lycopene is synthesised

    C   energy for the synthesis of lycopene comes from the breakdown of starch

    D   starch is broken down and lycopene is synthesised.

[END OF SECTION 1. NOW ATTEMPT THE QUESTIONS IN SECTION 2 OF YOUR QUESTION AND ANSWER BOOKLET]

National Qualifications
MODEL PAPER 1

Duration — 2 hours and 30 minutes

Biology
Section 1 — Answer Grid and Section 2

Fill in these boxes and read what is printed below.

Full name of centre

Town

Forename(s)

Surname

Number of seat

Date of birth

Day	Month	Year
D D	M M	Y Y

Scottish candidate number

Total marks — 100

SECTION 1 — 20 marks

Attempt ALL questions.

Instructions for completion of Section 1 are given on *Page two*.

SECTION 2 — 80 marks

Attempt ALL questions.

Write your answers clearly in the spaces provided in this booklet. Additional space for answers and rough work is provided at the end of this booklet. If you use this space you must clearly identify the question number you are attempting. Any rough work must be written in this booklet. You should score through your rough work when you have written your final copy.

Use **blue** or **black** ink.

Before leaving the examination room you must give this booklet to the Invigilator; if you do not you may lose all the marks for this paper.

HODDER GIBSON
LEARN MORE

SECTION 1— 20 marks

The questions for Section 1 are contained on *Page 51* — Questions.
Read these and record your answers on the answer grid on *Page 61* opposite.
Do NOT use gel pens.

1. The answer to each question is **either** A, B, C or D. Decide what your answer is, then fill in the appropriate bubble (see sample question below).

2. There is **only one correct** answer to each question.

3. Any rough working should be done on the additional space for answers and rough work at the end of this booklet.

Sample Question

The thigh bone is called the

    A   humerus

    B   femur

    C   tibia

    D   fibula.

The correct answer is **B**—femur. The answer **B** bubble has been clearly filled in (see below).

Changing an answer

If you decide to change your answer, cancel your first answer by putting a cross through it (see below) and fill in the answer you want. The answer below has been changed to **D**.

If you then decide to change back to an answer you have already scored out, put a tick (✓) to the **right** of the answer you want, as shown below:

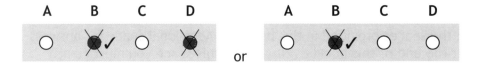

                                     or

SECTION 1 — Answer Grid

	A	B	C	D
1	○	○	○	○
2	○	○	○	○
3	○	○	○	○
4	○	○	○	○
5	○	○	○	○
6	○	○	○	○
7	○	○	○	○
8	○	○	○	○
9	○	○	○	○
10	○	○	○	○
11	○	○	○	○
12	○	○	○	○
13	○	○	○	○
14	○	○	○	○
15	○	○	○	○
16	○	○	○	○
17	○	○	○	○
18	○	○	○	○
19	○	○	○	○
20	○	○	○	○

MARKS | DO NOT WRITE IN THIS MARGIN

SECTION 2 — 80 marks

Attempt ALL questions

1. DNA and RNA molecules are found in the cells of both prokaryotes and eukaryotes.

    (a) Give **two** structural differences between DNA and RNA molecules.   **2**

    1 _____

    _____

    2 _____

    _____

    (b) Describe the function of RNA polymerase in the synthesis of a primary RNA transcript.   **1**

    _____

    _____

    (c) In eukaryotic cells, mRNA is spliced after transcription.

    Describe what happens during RNA splicing.   **2**

    _____

    _____

    (d) Describe the role of mRNA in cells.   **1**

    _____

    _____

MARKS | DO NOT WRITE IN THIS MARGIN

2. The diagram below shows details of a molecule of transfer RNA (tRNA).

    An area of the diagram has been enlarged.

(a) Complete the enlarged section of the diagram by adding letters to represent the complementary bases.   **1**

(b) Use the letter W to label the position of the amino acid binding site on the diagram.   **1**

(c) Name part X and describe its importance in translation.   **2**

    Name _____

    Importance _____

    _____

    _____

MARKS | DO NOT WRITE IN THIS MARGIN

3. *Mytilis edulis* and *Mytilis trossulus* are two closely related species of mussel which evolved after populations of their common ancestor became separated by a geographical barrier.

(a) (i) Name the type of speciation involved in this case.   **1**

_____

(ii) Describe the role of the barrier in the build up of genetic sequence differences between the evolving species.   **2**

_____

_____

_____

_____

(b) *M. edulis* and *M. trossulus* interbreed in some regions of their ranges.

(i) Give the term applied to regions where the interbreeding of closely related species occurs.   **1**

_____

(ii) Describe the evidence which would be required to confirm that *M. edulis* and *M. trossulus* are separate species.   **1**

_____

_____

_____

MARKS | DO NOT WRITE IN THIS MARGIN

4. During respiration in yeast, hydrogen ions are released from glucose molecules.

In experiments, these ions can decolourise resazurin, a blue indicator solution. The faster the rate of respiration, the faster resazurin will decolourise.

In an experiment, the effect of temperature on the rate of respiration in yeast was measured using the following method.

- A water bath was set up at 20°C

- Four tubes containing the substances shown on the table below were placed in the water bath

Tube	Contents
1	5 cm^3 yeast suspension
2	5 cm^3 glucose solution
3	5 cm^3 blue resazurin solution
4	empty

- The tubes were left in the water bath for 10 minutes then their contents were mixed into tube 4 and the time taken for the resazurin to decolourise was measured

- The procedure was repeated at a range of temperatures.

The results are shown in the table below.

Temperature of water bath (°C)	Time for resazurin to decolourise (s)
20	120
25	90
30	60
35	45
40	30
45	90

4. **(continued)**

(a) On the grid below, plot a line graph of the time taken for resazurin to decolourise against the temperature of the water bath.

(Additional graph paper if required will be found on *Page twenty-three*.)

2

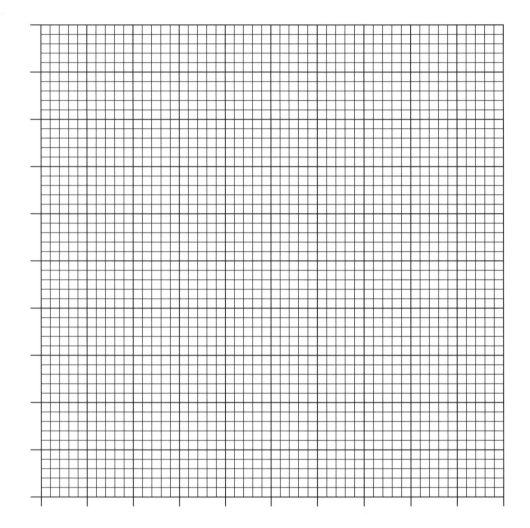

(b) (i) Identify the dependent variable in this experiment.

1

_____

(ii) Identify **two** variables which must be kept constant at each temperature to allow valid conclusions to be drawn from the results obtained.

1

1 _____

2 _____

MARKS | DO NOT WRITE IN THIS MARGIN

4. **(continued)**

(c) Describe the contents of a control tube which should be set up at each temperature to ensure that the decolourisation reaction was due to respiration.   1

_____

_____

_____

(d) Explain why the four tubes were kept in the water baths for 10 minutes before their contents were mixed.   1

_____

_____

_____

(e) Predict the time which would be taken for resazurin to decolourise if the experiment was repeated at 15°C.   1

_____ s

(f) In this experiment, hydrogen ions react with resazurin.

Describe how hydrogen ions cause the synthesis of ATP during respiration.   1

_____

_____

5. The following diagrams represent the structures of fish, amphibian and mammal hearts.

Heart R

Heart T

Heart S

direction of flow of blood

(a) Complete the table below by inserting the correct letters, animal group and type of circulation.

2

Heart	Animal group	Type of circulation
		single
	mammal	
S		

(b) Add arrows into vessels X, Y and Z **on the diagrams** to indicate the direction of blood flow within them.

2

(c) Explain how heart R is better adapted than heart S to allow high metabolic rates in the cells which it supplies with blood.

2

_____

_____

_____

MARKS | DO NOT WRITE IN THIS MARGIN

5. (continued)

(d) Give an example of a habitat with low oxygen availability and an adaptation found in an animal species which can live there.

Habitat _____   **1**

Adaptation _____

_____   **1**

6. Norway Spruce is an evergreen tree species which is adapted to survive adverse conditions of extreme cold in winter. The tree produces raffinose, a sugar which prevents needles from freezing solid during extreme temperature drops.

The table below shows the average raffinose content of samples of needles taken from trees over a period of seven months.

Month	Average raffinose content of needles (mg per g of needles)
June	0
July	1
August	2
September	3
October	9
November	30
December	50

(a) (i) Calculate the percentage increase in the average raffinose content of the needles between August and December.   **1**

Space for calculation

_____ %

MARKS | DO NOT WRITE IN THIS MARGIN

6. (a) (continued)

(ii) Describe how the changing raffinose content of the needles helps the Norway Spruce to survive.   **1**

_____

_____

(b) Norway spruce tolerates adverse conditions.

State how organisms can **avoid** adverse conditions.   **1**

_____

(c) Describe what is meant by the term "extremophile".   **1**

_____

_____

MARKS | DO NOT WRITE IN THIS MARGIN

7. The diagram below shows an industrial aerobic fermenter in which a culture of microorganisms is growing.

(a) State why fermenters such as this have steam pumped through them before adding the culture of microorganisms.

1

_____

_____

(b) Suggest why the rotating paddles are needed in the fermenter.

1

_____

_____

7.   (continued)

(c)   Complete the table below to show why the various conditions within the fermenter are maintained.

2

Condition	Reason for maintaining the condition
Aerobic environment	
Presence of glucose	
Constant pH	

8.   Papaya Ringspot Virus (PRSV) threatens food security because it damages papaya, a tropical fruit. Papaya plants with resistance to this virus have been produced using recombinant DNA technology as shown in the diagram below.

Step 1   Plasmid removed from the bacterium *Agrobacterium tumefaciens* and transformed with the PRSV resistance gene

Step 2   Modified plasmids returned to *Agrobacterium tumefaciens* cells

Step 3   Modified bacteria cultured then used to inoculate papaya cells and PRSV resistance genes taken into the genome of papaya cells

MARKS | DO NOT WRITE IN THIS MARGIN

8. (continued)

Step 4   Papaya cells grown on agar plate to develop resistant papaya plantlets

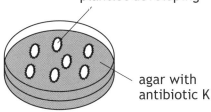

papaya
plantlet developing

agar with
antibiotic K

(a)   (i)   Name the enzyme used to seal the PRSV resistance gene into the plasmid in **Step 1**.   1

_____

(ii)   Explain why antibiotic K is added to the plate on which the papaya cells are grown in **Step 4**.   1

_____

_____

(iii)   Underline one word in each choice bracket to make the sentence which describes this example correct.   2

Agrobacterium is a $\left\{ \begin{array}{l} \text{prokaryote} \\ \text{eukaryote} \end{array} \right\}$ which has transferred genetic

sequences $\left\{ \begin{array}{l} \text{vertically} \\ \text{horizontally} \end{array} \right\}$ into the genome of a $\left\{ \begin{array}{l} \text{prokaryote} \\ \text{eukaryote} \end{array} \right\}$.

(b)   State why yeast cells are often used in recombinant DNA technology in preference to bacterial cells.   1

_____

_____

MARKS | DO NOT WRITE IN THIS MARGIN

9. Red spider mites *Tetranychus urticae* are pests which can threaten food security because they feed on crop plants such as maize, potatoes and strawberries.

Stages in the life cycle of the mite are shown in the diagram below.

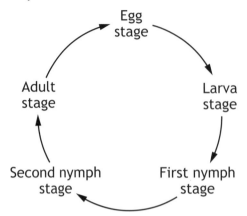

The **Chart** below shows the effects of temperature on the average duration of each stage in the life cycle and the **Table** shows how it affects features of egg laying in adult mites.

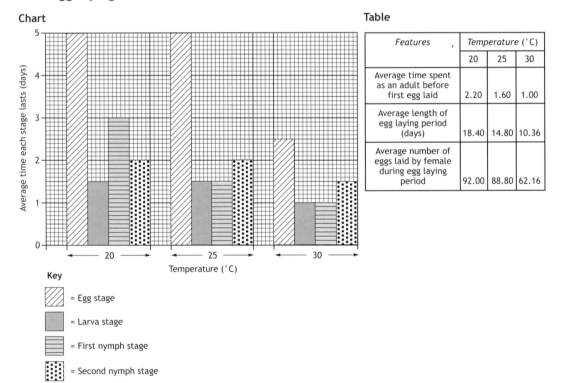

Table

Features	Temperature (°C)		
	20	25	30
Average time spent as an adult before first egg laid	2.20	1.60	1.00
Average length of egg laying period (days)	18.40	14.80	10.36
Average number of eggs laid by female during egg laying period	92.00	88.80	62.16

Key

◪ = Egg stage

▦ = Larva stage

▤ = First nymph stage

▨ = Second nymph stage

(a) From the **Chart**:

   (i) Identify the life cycle stage which is most affected by a rise in temperature from 20°C to 25°C. **1**

_____

MARKS | DO NOT WRITE IN THIS MARGIN

9. (a) (continued)

   (ii) Calculate the difference in average time of the egg stage when temperature is raised from 20°C to 30°C.   **1**

Space for calculation

_____days

(b) From the Table:

   (i) Describe how relationship affects.

      1   The length of egg-laying period;

      2   The average number of eggs laid in female red spider mites.

      1   Average length of egg laying period _____

    _____   **1**

      2   Average number of eggs laid per female _____

    _____   **1**

   (ii) Calculate the average number of eggs laid per female per day during the egg laying period at 20°C.

Space for calculation

Average number of eggs laid per female per day _____   **1**

(c) From the Chart and Table:

   (i) Decide whether each statement below is true or false and complete the grid by writing either **True** or **False** in each of the spaces provided.   **2**

Statement	True or False
Time for the first nymph stage is shortest at 30°C	
The egg stage lasts twice as long at 30°C as at 25°C	
Only the first nymph stage is affected by a change from 25°C to 30°C	
At 20°C some adults may take more than 2.2 days to start egg-laying	

   (ii) Calculate the average time that it takes for a female red spider mite to go through complete development from the beginning of her egg stage until she lays the first of her own eggs at 25°C.   **1**

Space for calculation

_____ days

MARKS | DO NOT WRITE IN THIS MARGIN

10. Answer **either A or B.**

    A   Describe the measurable components of biodiversity.   **5**

    OR

    B   Describe mass extinctions and their effect on biodiversity.   **5**

    Labelled diagrams may be used where appropriate.

MARKS | DO NOT WRITE IN THIS MARGIN

11. In an investigation into a sequence of metabolic reactions, a culture of the green alga *Chlorella* was kept in the dark for 24 hours. The culture was then exposed to light and samples of the algal cells obtained at four time intervals over 30 seconds of illumination. Substances present in extracts made from each sample were separated using chromatography techniques and the results are shown in the diagram below.

(a) Using information from the diagram, give the sequence in which the substances were produced.

1

_____ → _____ → _____ → _____

(b) The relative front value (Rf) of a substance can be calculated using the following formula:

$$Rf = \frac{\text{distance travelled by substance from origin}}{\text{distance travelled by solvent from origin}}$$

Use the formula to calculate the Rf value for alanine.

Rf value for alanine = _____

1

11. (continued)

(c) The graph below shows the relative concentrations of glyceraldehyde 3-phosphate (G3P) and RuBP in the algal cells when they were exposed to light and dark periods.

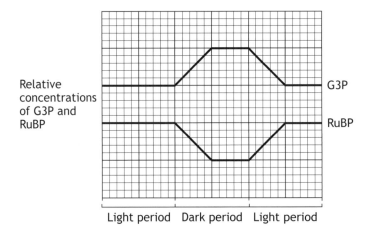

Relative concentrations of G3P and RuBP

G3P

RuBP

Light period   Dark period   Light period

(i) Identify evidence from the graph which supports the hypothesis that G3P can be converted to RuBP in light conditions.

1

_____

_____

(ii) During photosynthesis, G3P is produced when an intermediate substance is reduced by hydrogen and phosphorylated by ATP.

Name the intermediate substance involved and name the stage of photosynthesis which provides the hydrogen and ATP for this reaction.

Intermediate substance _____   1

_____

Stage _____   1

_____

MARKS | DO NOT WRITE IN THIS MARGIN

12. Integrated pest management (IPM) uses a combination of approaches to tackling crop pests.

(a) Complete the table below which shows different approaches and examples of their use.   3

Approach	Example of use
	applying insecticide to crops
Biological	
Cultural	

(b) Give an environmental problem associated with applying insecticide to crops.   1

_____

_____

13. The clownfish *Amphiprion ocellaris* lives among the stinging tentacles of the sea anemone *Heteractis magnifica*. The clownfish is highly territorial and drives off other fish which eat the anemone. The stinging tentacles of the anemone which are harmless to the clownfish help protect it from its own predators.

(a) (i) Name the type of symbiotic relationship described.   1

_____

(ii) Name one other type of symbiosis and describe its general features.   2

Name_____

Description_____

_____

MARKS | DO NOT WRITE IN THIS MARGIN

13. (continued)

(b) Worker honey bees are sterile. They are related to the queen bee and cooperate with other members of their colony to raise her offspring.

Explain the advantage to a worker of raising the offspring of the queen.   2

_____

_____

14. Answer **either A or B.**

A   Give an account of stem cells under the following headings:

   (i)   types of stem cell and their properties;   6

   (ii) therapeutic use of stem cells and ethical issues surrounding their use.   3

B   Give an account of genomic sequencing under the following headings:

   (i)   phylogenetics and molecular clocks;   6

   (ii) personal genomics and health.   3

Labelled diagrams may be used where appropriate.

[END OF MODEL PAPER]

ADDITIONAL SPACE FOR ANSWERS AND ROUGH WORK

ADDITIONAL GRAPH PAPER FOR QUESTION 4(a)

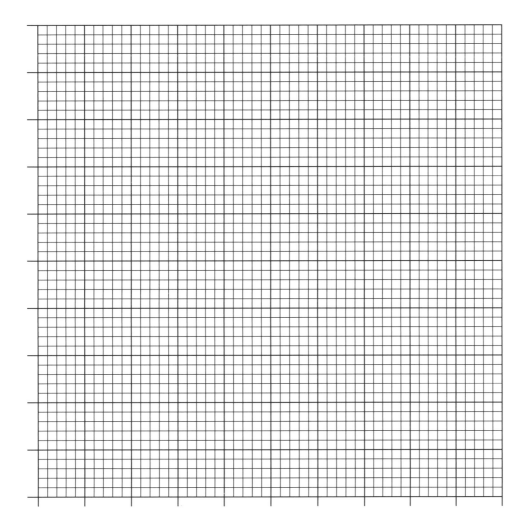

ADDITIONAL SPACE FOR ANSWERS AND ROUGH WORK

ADDITIONAL SPACE FOR ANSWERS AND ROUGH WORK

MARKS

DO NOT
WRITE IN
THIS
MARGIN

ADDITIONAL SPACE FOR ANSWERS AND ROUGH WORK

Model Paper 2

Whilst this Model Paper has been specially commissioned by Hodder Gibson for use as practice for the Higher (for Curriculum for Excellence) exams, the key reference documents remain the SQA Specimen Paper 2014 and SQA Past Paper 2015.

HODDER
GIBSON
LEARN MORE

National Qualifications
MODEL PAPER 2

Biology
Section 1—Questions

Duration — 2 hours and 30 minutes

Instructions for the completion of Section 1 are given on *Page two* of your question and answer booklet.

Record your answers on the answer grid on *Page three* of your question and answer booklet.

Before leaving the examination room you must give your question and answer booklet to the Invigilator; if you do not, you may lose all the marks for this paper.

SECTION 1 — 20 marks
Attempt ALL questions

1. The genetic material in the chloroplasts of eukaryotic plant cells is found in

    A   linear chromosomes

    B   circular plasmids

    C   circular chromosomes

    D   inner membranes.

2. A fragment of DNA was found to consist of 72 nucleotide base pairs.

    What is the total number of deoxyribose sugars in this fragment?

    A   24

    B   36

    C   72

    D   144

3. Which of the following statements related to cell differentiation is correct?

    A   Meristems are regions of differentiated cell types in plants.

    B   A differentiated cell only expresses genes that produce proteins characteristic of that cell type.

    C   Embryonic stem cells can differentiate into a limited range of cell types.

    D   Adult tissue stem cells can differentiate into all cell types.

4. Which of the following is true of polyploid plants?

    They have:

    A   reduced yield and the diploid chromosome number

    B   increased yield and the diploid chromosome number

    C   reduced yield and sets of chromosomes greater than diploid

    D   increased yield and sets of chromosomes greater than diploid.

5. Which of the following is **not** an example of a chromosome structure mutation?

    A   Substitution

    B   Duplication

    C   Translocation

    D   Inversion

6. The graph below shows the changes in number of human stem cells in a culture.

The activity of the enzyme glutaminase present in the cells over an eight day period is also shown.

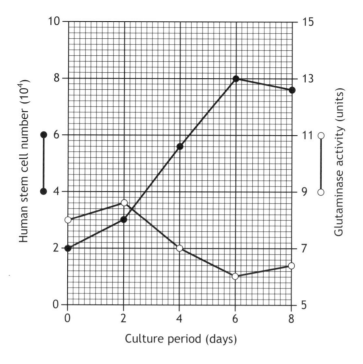

How many units of glutaminase activity were recorded when the cell number was 50% of its maximum over the eight day period?

A   3

B   4

C   8

D   9

7. Membranes can form small compartments within cells.

Small compartments have

A   high surface to volume ratios and allow low concentrations of substrates inside

B   high surface to volume ratios and allow high concentrations of substrates inside

C   low surface to volume ratios and allow low concentrations of substrates inside

D   low surface to volume ratios and allow high concentrations of substrates inside.

8. The reaction below is part of a metabolic pathway in cells.

    amino acids → polypeptides

    Which line in the table below correctly identifies the type of reaction and whether it releases or takes up energy?

	Type of reaction	Energy released or taken up
A	catabolic	released
B	anabolic	released
C	catabolic	taken up
D	anabolic	taken up

9. Using recombinant DNA technology, the bacterium *Escherichia coli* can be modified so that it can produce human insulin.

    The following steps are involved.

    1   Culture large quantities of *E. coli* in nutrient medium.
    2   Insert human insulin gene into *E. coli* plasmid DNA.
    3   Cut insulin gene from human chromosome using enzymes.
    4   Extract insulin from culture medium.

    The correct order of these steps is

    A   3, 2, 1, 4
    B   3, 1, 2, 4
    C   1, 4, 3, 2
    D   1, 2, 3, 4.

10. An investigation was carried out into the uptake of sodium ions by animal cells. The graph shows the rates of sodium ion uptake and breakdown of glucose at different concentrations of oxygen.

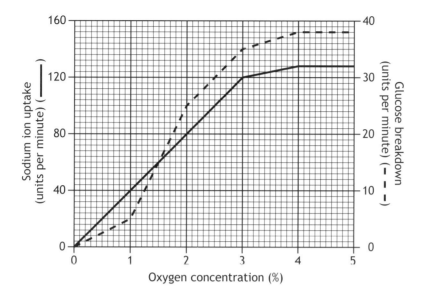

Calculate the number of units of sodium ions that are taken up over a 5 minute period when the concentration of oxygen in solution is 2%.

A 80

B 100

C 400

D 500

11. The diagram below shows a metabolic pathway that is controlled by end product inhibition.

For substance 4 to bring about end product inhibition of the overall pathway, with which of the following will it interact?

A Substance 1

B Substance 3

C Enzyme 1

D Enzyme 3

12. The giant bullfrog of South Africa lives in a habitat in which hot and dry conditions can occur at any time of the year. To survive these conditions, the frogs respond by becoming dormant.

   Which of the following descriptions applies to this type of dormancy?

   A   Predictive aestivation

   B   Consequential aestivation

   C   Predictive hibernation

   D   Consequential hibernation

13. Domestic chickens kept in social groups form a social hierarchy in which the most dominant birds give most pecks to others and receive least.

   In an investigation, four individual domestic chickens from a group were marked with lettered leg rings. The number of pecks given and received by each bird in one hour was recorded and the results shown in the table below.

		Number of pecks given by each bird to the others			
	Leg ring letter	W	X	Y	Z
Number of pecks received by each bird from the others	W	—	0	12	9
	X	15	—	8	7
	Y	0	0	—	0
	Z	0	0	10	—

   The social hierarchy among the chickens with the most dominant bird first is

   A   Y, Z, W, X

   B   X, W, Z, Y

   C   Y, W, Z, X

   D   X, Z, W, Y.

14. The action spectrum of photosynthesis is a measure of the ability of plants to

   A   absorb all wavelengths of light

   B   absorb light of different intensities

   C   extend the range of wavelengths absorbed

   D   use light of different wavelengths in photosynthesis.

15. The following statements refer to photosynthesis.

    1   Carbon dioxide is fixed by RuBisCO.

    2   Water is split into hydrogen and oxygen.

    3   G3P is used to regenerate RuBP.

    Which of the statements correctly refer to the Calvin cycle?

    A   1 and 2 only

    B   1 and 3 only

    C   2 and 3 only

    D   1, 2 and 3

16. Natural selection reduces in-breeding depression in self-pollinating plant species by eliminating

    A   deleterious alleles

    B   recessive alleles

    C   heterozygous alleles

    D   dominant alleles.

17. Northern elephant seals have very low genetic variation caused by a catastrophic decline in numbers of this species due to over-hunting by humans.

    Present day animals have all descended from the small number that survived.

    What term is used to refer to the loss of genetic variation associated with a serious decline in population?

    A   Founder effect

    B   Bottleneck effect

    C   Stabilising selection

    D   Directional selection

18. The list of statements below refers to advantages gained by cooperative hunting behaviour.

    1   Much larger prey may be killed than by hunting alone.

    2   Both dominant and subordinate animals benefit.

    3   Individuals gain more energy than by hunting alone.

    Which of the statements could be true of cooperative hunting?

    A   1 and 2 only

    B   1 and 3 only

    C   2 and 3 only

    D   1, 2 and 3

19. Which line in the table below best describes the effects of altruistic behaviour on the donor and the recipient?

	Effect on donor	Effect on recipient
A	benefits	benefits
B	benefits	harmed
C	harmed	benefits
D	harmed	harmed

20. A species of South American ant inhabits the thorns of a species of *Acacia*. The ant receives nectar and shelter from the plant. The ants protect the plant from attack by other insects.

This is an example of

A parasitism

B mutualism

C cooperative hunting

D predation.

National
Qualifications
MODEL PAPER 2

Biology
Section 1 — Answer Grid
and Section 2

Duration — 2 hours and 30 minutes

Fill in these boxes and read what is printed below.

Full name of centre

Town

Forename(s)

Surname

Number of seat

Date of birth

Day	Month	Year
D D	M M	Y Y

Scottish candidate number

Total marks — 100

SECTION 1 — 20 marks

Attempt ALL questions.

Instructions for completion of Section 1 are given on *Page two*.

SECTION 2 — 80 marks

Attempt ALL questions.

Write your answers clearly in the spaces provided in this booklet. Additional space for answers and rough work is provided at the end of this booklet. If you use this space you must clearly identify the question number you are attempting. Any rough work must be written in this booklet. You should score through your rough work when you have written your final copy.

Use **blue** or **black** ink.

Before leaving the examination room you must give this booklet to the Invigilator; if you do not you may lose all the marks for this paper.

SECTION 1— 20 marks

The questions for Section 1 are contained on *Page 87* — Questions.
Read these and record your answers on the answer grid on *Page 97* opposite.
Do NOT use gel pens.

1. The answer to each question is **either** A, B, C or D. Decide what your answer is, then fill in the appropriate bubble (see sample question below).

2. There is **only one correct** answer to each question.

3. Any rough working should be done on the additional space for answers and rough work at the end of this booklet.

Sample Question

The thigh bone is called the

    A   humerus

    B   femur

    C   tibia

    D   fibula.

The correct answer is **B**—femur. The answer **B** bubble has been clearly filled in (see below).

Changing an answer

If you decide to change your answer, cancel your first answer by putting a cross through it (see below) and fill in the answer you want. The answer below has been changed to **D**.

If you then decide to change back to an answer you have already scored out, put a tick (✓) to the **right** of the answer you want, as shown below:

SECTION 1 — Answer Grid

	A	B	C	D
1	○	○	○	○
2	○	○	○	○
3	○	○	○	○
4	○	○	○	○
5	○	○	○	○
6	○	○	○	○
7	○	○	○	○
8	○	○	○	○
9	○	○	○	○
10	○	○	○	○
11	○	○	○	○
12	○	○	○	○
13	○	○	○	○
14	○	○	○	○
15	○	○	○	○
16	○	○	○	○
17	○	○	○	○
18	○	○	○	○
19	○	○	○	○
20	○	○	○	○

MARKS | DO NOT WRITE IN THIS MARGIN

SECTION 2 — 80 marks

Attempt ALL questions

1. The polymerase chain reaction (PCR) can be used to make many copies of a small sample of DNA.

   (a) Describe how the strands of DNA are separated during PCR. **1**

   _____

   _____

   (b) Explain why primers are required in PCR. **1**

   _____

   _____

   (c) Name the enzyme which replicates the section of DNA to be amplified in PCR. **1**

   _____

   (d) Calculate how many molecules of DNA would be produced after a single molecule of DNA went through four complete cycles of PCR. **1**

   Space for calculation

   _____ molecules

MARKS

2. Split genes are genes which have the coding DNA split by sections of DNA which do not code for protein.

The diagram below shows the synthesis of insulin from a split gene.

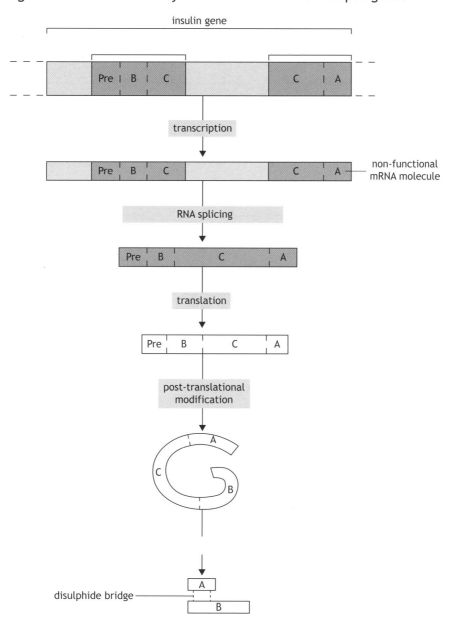

(a) Name the non-coding regions of eukaryotic genes. 1

(b) Name the non-functional mRNA molecule which contains both coding and non-coding regions. 1

2. (continued)

(c) Explain the significance of alternative RNA splicing in terms of gene expression.

1

_____

_____

(d) Post-translational modification of the pre-insulin polypeptide involves cutting and combining chains.

Give **one** other way in which a polypeptide can be modified following translation.

1

_____

3. Severe combined immunodeficiency (SCID) is a rare inherited condition.

Children with SCID have white blood cells that lack the functional gene that codes for the enzyme adenosine deaminase (ADA). Without this enzyme, toxins accumulate and destroy the white blood cells, made in bone marrow and which normally fight infection.

Gene therapy using bone marrow stem cells has been used in the treatment of some children with this condition as shown below.

Step 1   gene for ADA isolated from healthy human cells

↓

Step 2   ADA gene inserted into a virus

↓

Step 3   bone marrow stem cells isolated from child and infected with the modified virus

↓

Step 4   modified stem cells injected back to child

3.   (continued)

(a)   Give **two** properties of stem cells which makes them suitable for use in this therapy.

1 _____   1

_____

2 _____   1

_____

(b)   The graph below shows the number of functional white blood cells in a patient who had undergone this treatment. Unaffected children have a white blood cell count which ranges between 5000 and 8000 per mm³ of blood.

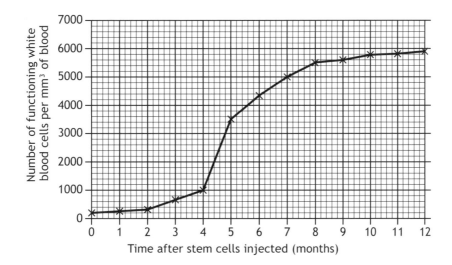

(i)   Calculate the percentage increase in the number of functional white blood cells per mm³ of blood between the fourth and fifth month after the stem cells had been injected.   1

Space for calculation

_____ %

MARKS

3. **(b) (continued)**

    (ii) Calculate the average increase per month in the number of functional white blood cells per mm³ of blood over the 12 months.   **1**

    Space for calculation

_____

    (iii) State the number of months after stem cells were injected that the white blood cell count reached the lower value of an unaffected child.   **1**

    _____ months

(c) Give **one** use of stem cells in medical research.   **1**

_____

_____

4. The phylogenetic tree below shows the evolutionary relationship between some mammals.

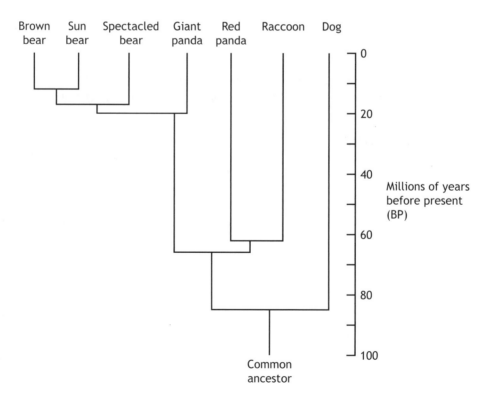

MARKS | DO NOT WRITE IN THIS MARGIN

4. **(continued)**

(a) Give **one** source of data which can be used to confirm the evolutionary relationship between the mammals shown on the diagram.   1

_____

(b) Calculate how many million years separated the divergence of dogs and other species from the divergence of brown bears and sun bears.   1

Space for calculation

_____ million years BP

(c) State how long ago the last common ancestor of the spectacled bear and the giant panda existed.   1

_____ million years BP

(d) State the number of other species with which the raccoon shared a common ancestor 66 million years before present.   1

_____ species

MARKS | DO NOT WRITE IN THIS MARGIN

5. Answer **either A or B.**

   A   Give an account of the formation and maintenance of hybrid zones   4

   OR

   B   Give an account of the role of genetic drift in evolution.   4

   Labelled diagrams may be used where appropriate.

MARKS | DO NOT WRITE IN THIS MARGIN

6. During pharmaceutical trials for a new drug, a healthy subject volunteered to drink 1 litre of water.

Samples of urine were taken from the subject over a 180 minute period. These samples were used to determine the rate of urine production and the salt concentration of the urine.

The results are shown in **Graph 1** below.

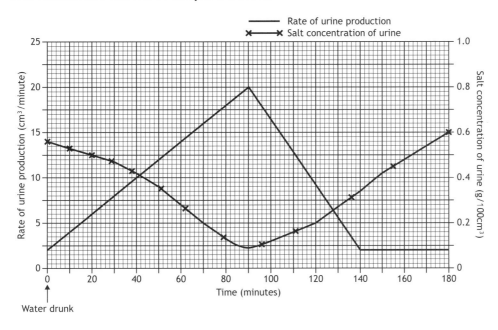

(a) From **Graph 1**, give the time taken, after drinking the water, for the rate of urine production to return to the starting value.

1

_____ minutes

(b) Use values from **Graph 1** to describe the changes in the rate of urine production over the 3 hour period.

2

_____

_____

_____

_____

MARKS | DO NOT WRITE IN THIS MARGIN

6. (continued)

(c) From **Graph 1**, identify the time period during which the greatest decrease in salt concentration in the urine occurred.   1

Tick (✓) the correct box below.

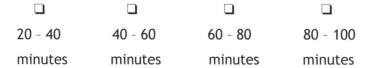

❏   ❏   ❏   ❏

20 – 40   40 – 60   60 – 80   80 – 100

minutes   minutes   minutes   minutes

(d) In a further investigation, 10 mg of the new drug under trial was administered to the volunteer who then drank a further 1 litre of water.

The rate of urine production was measured and the results are shown in **Graph 2** below.

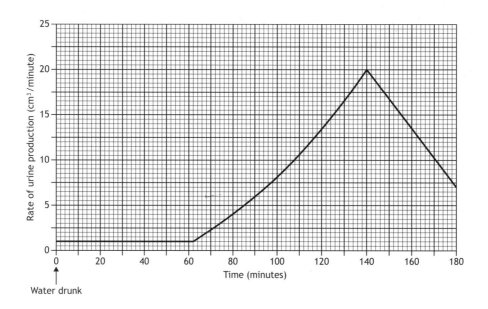

From **Graphs 1 and 2**, calculate the difference the drug has made to the rate of production of urine 90 minutes into the trial.   1

Space for calculation

_____ cm³/minute

MARKS | DO NOT WRITE IN THIS MARGIN

6. (continued)

    (e) Describe the type of analysis which would provide information about the likelihood of success before the drug could be used as part of personalised medical treatment of a specific patient.   1

    _____

    _____

7. Catechol oxidase is an enzyme found in apple tissue. It is involved in the reaction which produces the brown pigment that forms in cut or damaged apples.

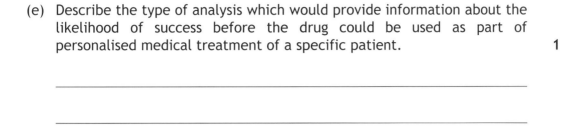

catechol
(colourless substance
in apple tissue)
$\xrightarrow{\text{catechol oxidase}}$
brown pigments

The effect of the concentration of lead ethanoate on this reaction was investigated.

10 g of apple tissue was cut up, added to 10 cm³ of distilled water and then liquidised and filtered. This produced an extract containing both catechol and catechol oxidase.

Test tubes were set up as described in **Table 1** and kept at 20°C in a water bath.

Table 1

Test tube	Contents of test tubes
A	sample of extract + 1cm³ 0.01% lead ethanoate solution
B	sample of extract + 1cm³ 0.1% lead ethanoate solution

Every 10 minutes, the tubes were placed in a colorimeter which measured how much brown pigment was present.

The more brown pigment present the higher the colorimeter reading.

7. **(continued)**

The results are shown in **Table 2.**

Table 2

Time (minutes)	Colorimeter reading (units)	
	Test tube A	Test tube B
	sample of extract + 0.01% lead ethanoate	sample of extract + 0.1% lead ethanoate
0	1.8	1.6
10	5.0	2.0
20	6.0	2.2
30	6.4	2.4
40	7.0	2.4
50	7.6	2.4
60	7.6	2.4

(a) Identify **two** variables not already mentioned that would have to be kept constant.

1 _____ 1

2 _____ 1

(b) Explain why the initial colorimeter readings were not 0.0 units.   1

_____

_____

(c) The results for the extract with 0.1% lead ethanoate are shown in the graph below.

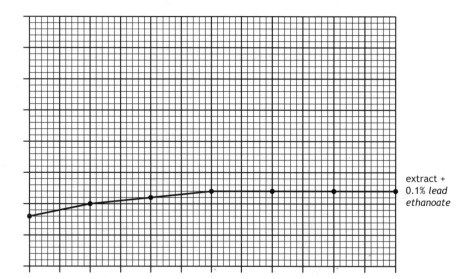

extract + 0.1% lead ethanoate

MARKS | DO NOT WRITE IN THIS MARGIN

7. (c) (continued)

Use information from **Table 2** to complete the graph by:

    (i) adding the scale and label to each axis;   **1**

    (ii) presenting the results for the extract + 0.01% lead ethanoate solution and labelling the line.   **1**

(Additional graph paper, if required, will be found on *Page twenty-five*)

(d) State the effect of the concentration of lead ethanoate solution on the activity of catechol oxidase.   **1**

_____

_____

(e) Describe a suitable control for this investigation.   **1**

_____

_____

(f) State how the procedure could be improved to increase the reliability of the results.   **1**

_____

_____

8. The diagram below represents a section through the inner mitochondrial membrane and some of the processes leading to ATP formation.

8. **(continued)**

(a) Name the two hydrogen carriers that link the citric acid cycle to the electron transfer chain.   1

1 _____

2 _____

(b) State the role of the electrons that pass down the electron transfer chain.   1

_____

_____

(c) Name protein X, responsible for the regeneration of ATP.   1

_____

(d) State the role of oxygen in the electron transfer chain.   1

_____

_____

9. The diagram below represents a vertical section through the skin of a mammal.

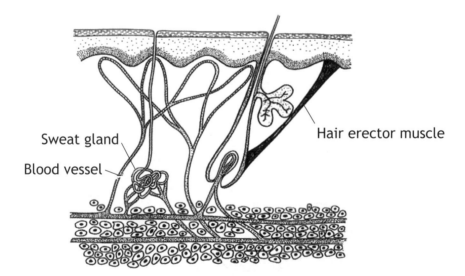

Sweat gland

Blood vessel

Hair erector muscle

MARKS | DO NOT WRITE IN THIS MARGIN

9. **(continued)**

(a) Select **one** structure labelled in the diagram and explain its role in response to a decrease in temperature.

Structure _____

Role _____

_____ 1

(b) Explain the importance of regulating body temperature to the metabolism of humans.

_____

_____

_____ 2

(c) Give the term used to describe organisms whose internal environment is dependent upon their external environment.

_____ 1

10. Plasmids are often used as vectors in recombinant DNA technology.

Many different restriction endonuclease enzymes are used in recombinant DNA technology. Each enzyme cuts the DNA of the plasmid at a specific base sequence called its restriction site.

The diagram below shows the position of four different restriction sites, P, Q, R and S. The distances between the restriction sites are measured in kilobases (kb) of DNA.

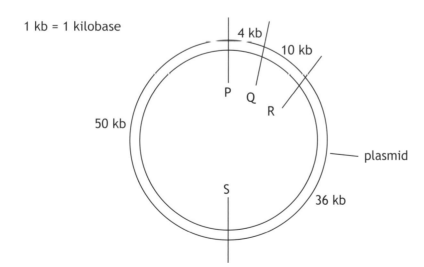

MARKS | DO NOT WRITE IN THIS MARGIN

10. (continued)

Two restriction endonuclease enzymes were used to cut the plasmid DNA.

The sections of plasmid DNA (fragments) which resulted were separated by a process called gel electrophoresis.

The positions of the fragments in the electrophoresis gel are shown in the chart below.

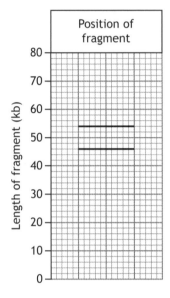

(a) From the information in the **chart** and **diagram** state which restriction sites were cut.

1

_____ and _____

(b) Describe the role of a vector in recombinant DNA technology.

1

_____

_____

(c) Name the enzyme which can be used to seal the foreign DNA into the plasmid.

1

_____

(d) State one advantage of using bacterial cells, such as *E. coli*, as the recipient for foreign DNA.

1

_____

_____

MARKS | DO NOT WRITE IN THIS MARGIN

10. (continued)

(e) Describe a problem associated with expressing animal DNA in *E. coli* and name an alternative organism which can be used to overcome this problem.

Problem _____ 1

_____

Name _____ 1

11. The graph below shows the growth of a population of a bacterial species in a fermenter over a 60 hour period.

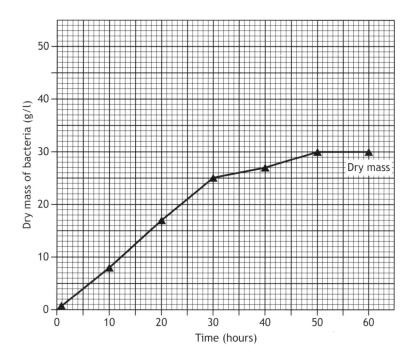

(a) Explain why the fermenter would be sterilised at the start of the procedure. 1

_____

_____

(b) State **two** other factors that are controlled to provide optimum culture conditions for growth of the bacteria. 1

1 _____

2 _____

MARKS | DO NOT WRITE IN THIS MARGIN

11. (continued)

(c) Calculate the average rate of increase in dry mass of the bacteria during the culture period.

Space for calculation

_____ g/l/hour

1

(d) Secondary metabolism can confer an ecological advantage to a microorganism by producing substances not associated with growth.

Give **one** example of a secondary metabolite and explain how it can confer an ecological advantage to the microorganism which produces it.

Example _____   1

Explanation _____

_____   1

12. Fire ants *Solenopsis invicta* were accidentally brought to the United States by ships from South America during the 1930s. The imported fire ants were more competitive than native ant species and quickly displaced them.

(a) (i) Give the name applied to introduced species which have become established in wild communities.   1

_____

(ii) Fire ants are now classified as invasive species in the United States.

Give **one** reason why the population of an invasive species may increase at the expense of native species.   1

_____

_____

MARKS | DO NOT WRITE IN THIS MARGIN

12. (continued)

(b) Species of *phorid* flies attack fire ants in South America.

Female *phorid* flies lay eggs in the ant's body. When the eggs hatch, the larva burrows into the head of the fire ant, and feeds on its internal tissues.

An investigation was carried out into the control of fire ant using an integrated pest management (IPM) approach which combined the use of *phorid* flies and an insecticide.

The graph below shows the results.

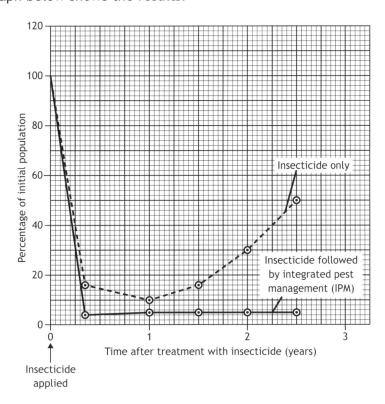

(i) Compare the effect of using the insecticide only with the integrated approach using insecticide and *phorid* flies.   2

_____

_____

(ii) Suggest how natural selection might account for the rise in the fire-ant population which occurred with the insecticide-only treatment after year 1.   2

_____

_____

_____

MARKS | DO NOT WRITE IN THIS MARGIN

12. (b)   (continued)

(iii) Describe **one** possible risk of introducing the *phorid* flies to control fire-ant populations in the United States.

1

_____

_____

(iv) Explain how the information about the *phorid* flies confirms that they have a parasitic relationship with the fire ants.

1

_____

_____

13. Pigs are sometimes reared in intensive units in which temperature is controlled. Agricultural scientists investigated the effect of temperature on growth rate and the efficiency with which the pigs converted food to biomass.

The table below shows the results of this investigation.

Temperature (°C)	Mean growth rate (g/day)	Conversion efficiency of food to biomass (%)
0	540	19
10	800	42
20	850	48
30	450	37
35	310	37

(a) Explain why pigs of the same breed, with similar genotypes were used in this investigation.

1

_____

_____

(b)   (i) Describe the effect of temperature on mean growth rate.

2

_____

_____

_____

MARKS | DO NOT WRITE IN THIS MARGIN

13. (b) (continued)

(ii) Express the mean growth rate at 30°C to 0°C as the simplest whole number ratio.

Space for calculation

1

_____ : _____

30°C   0°C

(iii) It was concluded from these data that the mean growth rate of the pigs was fastest at 20°C.

Describe what would need to be done to improve the validity of this conclusion.

1

_____

_____

(iv) The efficiency of conversion of food to biomass is lower at 0°C than it is at 20°C.

Suggest an explanation for the lower efficiency.

1

_____

_____

(v) Give **one** example of a type of behaviour which could indicate poor welfare in domesticated pigs.

1

_____

MARKS

14. Answer **either A or B.**

    A   Write notes on biodiversity under the following headings:

       (i) measuring biodiversity;   4

      (ii) threats to biodiversity.   5

OR

    B   Write notes on human food supply under the following headings:

       (i) food security and population;   3

      (ii) factors affecting food production.   6

Labelled diagrams may be used where appropriate.

[END OF MODEL PAPER]

ADDITIONAL SPACE FOR ANSWERS AND ROUGH WORK

ADDITIONAL GRAPH PAPER FOR QUESTION 7 (c)

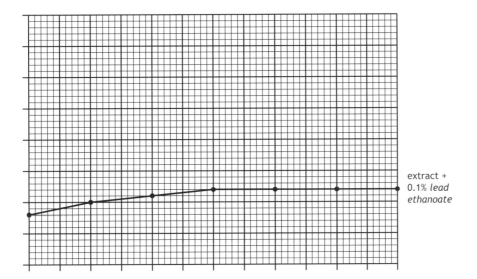

extract + 0.1% *lead ethanoate*

ADDITIONAL SPACE FOR ANSWERS AND ROUGH WORK

ADDITIONAL SPACE FOR ANSWERS AND ROUGH WORK

MARKS | DO NOT WRITE IN THIS MARGIN

ADDITIONAL SPACE FOR ANSWERS AND ROUGH WORK

MARKS

Model Paper 3

Whilst this Model Paper has been specially commissioned by Hodder Gibson for use as practice for the Higher (for Curriculum for Excellence) exams, the key reference documents remain the SQA Specimen Paper 2014 and SQA Past Paper 2015.

National
Qualifications
MODEL PAPER 3

Biology
Section 1—Questions

Duration — 2 hours and 30 minutes

Instructions for the completion of Section 1 are given on *Page two* of your question and answer booklet.

Record your answers on the answer grid on *Page three* of your question and answer booklet.

Before leaving the examination room you must give your question and answer booklet to the Invigilator; if you do not, you may lose all the marks for this paper.

SECTION 1 — 20 marks

Attempt ALL questions

1. The statements below refer to events in the replication of a lagging strand of DNA.

   1 primers bind to template chains

   2 DNA ligase forms sugar-phosphate links

   3 hydrogen bonds break

   4 DNA polymerase adds free nucleotides to strand

   5 double helix unwinds

   In which order do these events occur?

   A 1, 2, 4, 3, 5

   B 5, 3, 4, 1, 2

   C 5, 3, 1, 4, 2

   D 1, 3, 2, 4, 5

2. After a few seconds in a thermal cycling PCR machine, a DNA sequence has been amplified to give 64 copies.

   How many more cycles of PCR would be needed to amplify the sequence to 2048 copies?

   A 4

   B 5

   C 6

   D 11

3. The list below shows different ribonucleic acid molecules which occur in living cells.

   1 mRNA

   2 tRNA

   3 rRNA

3. **(continued)**

Which line in the table below correctly matches the types of RNA with their functions?

	RNA functions		
	Picks up and carries specific amino acids	*Carries copy of the genetic codes*	*Combines with protein to form ribosomes*
A	2	1	3
B	1	3	2
C	2	3	1
D	3	1	2

4. A frame-shift mutation can be the result of

A   a nucleotide deletion

B   an expansion of a nucleotide sequence repeat

C   a single nucleotide substitution

D   a nucleotide change at a splice site.

5. The table below shows information about ancestral and modern *Brassica* crops.

The modern species have been produced by hybridisation of two ancestral species followed by a doubling of the chromosome number in the hybrids.

Brassica species	*Ancestral or Modern*	*Crop*	*Chromosomes*
B. oleracea	ancestral	cabbage	2n = 18
B. nigra	ancestral	black mustard	2n = 16
B. rapa	ancestral	turnip	2n = 20
B. juncea	modern	Indian mustard	2n = 36
B. carinata	modern	Ethiopean mustard	2n = 34
B. napus	modern	oilseed rape	2n = 38

Which of the following shows the correct ancestral hybridisation and the modern species produced?

A   turnip x cabbage → Indian mustard

B   turnip x black mustard → Ethiopean mustard

C   cabbage x black mustard → Indian mustard

D   cabbage x turnip → oilseed rape

Page three

6. The diagram below shows a bacterial cell which has been magnified 600 times.

2.4 mm

What is the actual length of the bacterial cell?

A 0.4 μm

B 4.0 μm

C 40 μm

D 0.4 mm

7. The list below shows types of reaction involved in respiration.

1 phosphorylation

2 dehydrogenation

3 fermentation

Which line in the table below correctly identifies the types of reaction occurring at the stages in respiration given?

	Stages in respiration		
	Pyruvate converted to lactate	Glucose converted to intermediates in glycolysis	Citrate converted to oxaloacetate
A	1 and 2	2 only	2 and 3
B	3 only	1 only	2 only
C	3 only	1 and 2	1 and 2
D	1 and	3 only	1 and 2

8. The figure below shows how the ATPase activity of the sodium-potassium pump is affected by the concentrations of sodium and potassium ions.

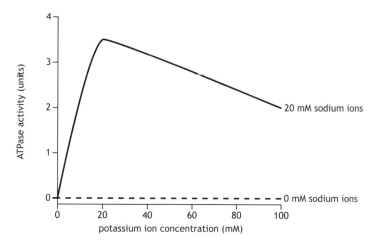

What valid conclusion can be drawn from this information?

A   ATPase activity requires the presence of both sodium and potassium ions.

B   The optimum concentration of sodium ions for ATPase activity is 20 mM.

C   The presence of potassium ions inhibits ATPase activity.

D   ATPase activity requires the presence of sodium ions only.

9. The diagram below shows a bacterial plasmid that carries genes for resistance to the antibiotics ampicillin and tetracycline.

Pst I and *Bam* HI are restriction sites in the plasmid.

Which line in the table below correctly identifies the antibiotics to which bacteria transformed to contain the plasmid would be resistant, if a new gene were inserted into the restriction sites as shown?

	Restricted site into which new gene is inserted	Antibiotic(s) to which transformed bacteria would be resistant
A	*Pst* I	ampicillin
B	*Pst* I	tetracycline and ampicillin
C	*Bam* HI	tetracycline and ampicillin
D	*Bam* HI	ampicillin

10. Four humans exercised on a treadmill for 10 minutes and their maximum rate of oxygen consumption measured.

Which of the four individuals would be considered least fit?

Individual	Body mass (kg)	Maximum oxygen consumption (litres per minute)
A	70	2.35
B	75	2.72
C	80	3.08
D	85	3.26

11. The diagram below represents a vertebrate heart.

Which line in the table identifies correctly the class of vertebrate with hearts of this type and the type of circulation involved?

	Vertebrate class	Type of circulation
A	fish	double
B	fish	single
C	amphibian	double
D	amphibian	single

12. Which of the following statements refers to regulators?

A   Their internal environment changes with external environment.

B   They live only within a narrow ecological niche.

C   The maintenance of their internal environment has a high metabolic cost.

D   Their optimum metabolic rate is maintained by behavioural responses alone.

13. Liver tissue contains an enzyme involved in the breakdown of alcohol. The graph below shows the effect of different concentrations of copper ions on the breakdown of alcohol by this enzyme over a 30 minute period.

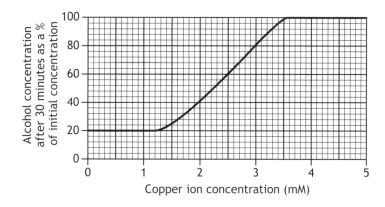

Which of the following conclusions can be drawn from the graph?

A   4.5mM copper ions has no effect on the enzyme activity.

B   2.5mM copper ions halves the enzyme activity.

C   0.5mM copper ions completely inhibits enzyme activity.

D   Enzyme activity increases when copper ion concentration is increased from 1mM to 2mM.

14. The diagram below shows a branched metabolic pathway.

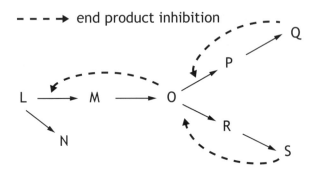

Which reaction would occur if both Q and S were present in the cell in high concentrations?

A   L → M

B   R → S

C   O → P

D   L → N

15. In plant productivity, the economic yield is the

A   total plant biomass

B   mass of the desired product

C   increase in mass due to photosynthesis minus loss due to respiration

D   rate of generation of new biomass per unit area per unit of time.

16. The greater honeyguide *Indicator indicator* is a bird which feeds on beeswax, bee eggs and grubs but is unable to open hives to feed. These birds lead humans to beehives they find and wait while the humans open the hives and remove the honey-containing combs. The birds then feed on material from the discarded hives.

Which feeding relationship does this example show?

A   parasitism

B   predation

C   competition

D   mutualism

17. The graph below shows the effect of adding different levels of fertiliser on the yield of a crop plant.

The percentage increase in yield obtained when the fertiliser level is increased from 40 to 80 kg ha^{-1} is

A   24

B   40

C   58

D   75

18. A species whose role is vital for the survival of many other species in an ecosystem is called a

A   native species

B   indicator species

C   keystone species

D   naturalised species

19. Which of the following does **not** cause loss of biodiversity?

A   habitat fragmentation

B   the bottleneck effect

C   intraspecific competition

D   overexploitation

20. Which line in the table below shows correctly how biological factors change after a mass extinction event?

	Biological factors		
	Speciation rate	*Occupied niches*	*Competition*
A	lower	more	more
B	lower	fewer	more
C	higher	more	less
D	higher	fewer	less

National
Qualifications
MODEL PAPER 3

Biology
Section 1 — Answer Grid
and Section 2

Duration — 2 hours and 30 minutes

Fill in these boxes and read what is printed below.

Full name of centre

Town

Forename(s)

Surname

Number of seat

Date of birth

Day	Month	Year
D D	M M	Y Y

Scottish candidate number

Total marks — 100

SECTION 1 — 20 marks

Attempt ALL questions.

Instructions for completion of Section 1 are given on *Page two*.

SECTION 2 — 80 marks

Attempt ALL questions.

Write your answers clearly in the spaces provided in this booklet. Additional space for answers and rough work is provided at the end of this booklet. If you use this space you must clearly identify the question number you are attempting. Any rough work must be written in this booklet. You should score through your rough work when you have written your final copy.

Use **blue** or **black** ink.

Before leaving the examination room you must give this booklet to the Invigilator; if you do not you may lose all the marks for this paper.

SECTION 1— 20 marks

The questions for Section 1 are contained one *Page 125 —* Questions. Read these and record your answers on the answer grid on *Page 137* opposite.
Do NOT use gel pens.

   1.   The answer to each question is **either** A, B, C or D. Decide what your answer is, then fill in the appropriate bubble (see sample question below).

   2.   There is **only one correct** answer to each question.

   3.   Any rough working should be done on the additional space for answers and rough work at the end of this booklet.

Sample Question

The thigh bone is called the

     A   humerus

     B   femur

     C   tibia

     D   fibula.

The correct answer is **B**—femur. The answer **B** bubble has been clearly filled in (see below).

Changing an answer

If you decide to change your answer, cancel your first answer by putting a cross through it (see below) and fill in the answer you want. The answer below has been changed to **D**.

If you then decide to change back to an answer you have already scored out, put a tick (✓) to the **right** of the answer you want, as shown below:

SECTION 1 — Answer Grid

	A	B	C	D
1	○	○	○	○
2	○	○	○	○
3	○	○	○	○
4	○	○	○	○
5	○	○	○	○
6	○	○	○	○
7	○	○	○	○
8	○	○	○	○
9	○	○	○	○
10	○	○	○	○
11	○	○	○	○
12	○	○	○	○
13	○	○	○	○
14	○	○	○	○
15	○	○	○	○
16	○	○	○	○
17	○	○	○	○
18	○	○	○	○
19	○	○	○	○
20	○	○	○	○

MARKS | DO NOT WRITE IN THIS MARGIN

SECTION 2 — 80 marks

Attempt ALL questions

1. The diagram below shows part of a DNA template strand and part of a primary RNA transcript synthesised from it.

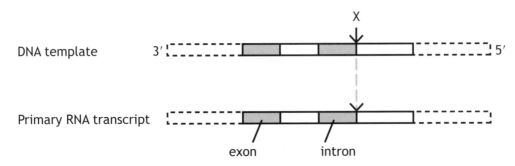

(a) Give the term used to describe the production of a primary RNA transcript from a DNA template.   **1**

_____

(b) DNA is encoded in triplet sequences.

Explain what is meant by this.   **1**

_____

_____

(c) Using information in the diagram:

   (i) explain what is meant by the term "antiparallel";   **1**

_____

_____

   (ii) describe how splicing can change the structure of the primary RNA transcript;   **1**

_____

_____

MARKS | DO NOT WRITE IN THIS MARGIN

1. (c) (continued)

(iii) describe a possible effect on the primary RNA transcript of a single nucleotide mutation at point X on the DNA strand.

1

_____

_____

2. The diagram below represents some of the amino acids linked together in an enzyme molecule.

Amino acid

(a) Name the bonds which connect the amino acids to form the chain.

1

_____

(b) Name **one** type of interaction which can hold the amino acid chain in folds as shown at X.

1

_____

(c) Explain why it is important that enzyme molecules are held in specific shapes.

1

_____

_____

(d) Explain the term "induced fit" as applied to enzyme action.

1

_____

_____

MARKS | DO NOT WRITE IN THIS MARGIN

3. The diagram below shows a phylogenetic tree for six animal groups in existence today.

Also shown are features which arose as the various groups evolved.

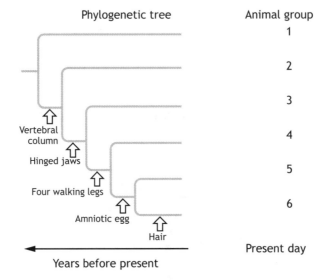

Phylogenetic tree   Animal group
1
2
3
4
5
6

Vertebral column
Hinged jaws
Four walking legs
Amniotic egg
Hair

Present day

Years before present

(a) Give the animal group(s) which do(es) not have hinged jaws. **1**

_____

(b) List the features of animal group 4. **1**

_____

(c) Give the animal group most closely related to group 3. **1**

_____

(d) State **two** sources of evidence which allow phylogenetic trees to be constructed. **2**

1 _____

2 _____

MARKS | DO NOT WRITE IN THIS MARGIN

4. In an experiment, radio-active amino acids were added to a culture of human liver cells.

Every minute for 5 minutes a sample of cells was withdrawn and killed. The cells were broken up, the ribosomes isolated and the mass of amino acids associated with the ribosomes measured by determining their radioactivity as shown in the table below.

Time (mins)	Mass of amino acids associated with ribosomes (Radioactivity units)
0	0
1	40
2	75
3	115
4	120
5	120

(a) Suggest the hypothesis which was being tested by this experiment. 1

_____

_____

(b) **Using values from the table,** describe how the mass of amino acids associated with the ribosomes changes over the 5 minute period of the experiment. 2

_____

_____

(c) Predict the mass of amino acids which would be expected to have reached the ribosomes after 6 minutes. 1

_____ radioactivity units

MARKS | DO NOT WRITE IN THIS MARGIN

5. The grid below refers to materials involved in the electron transport chain of aerobic respiration.

A ATP	B NAD	C water	D oxygen
E ATP synthase	F hydrogen ions	G electrons	H FAD

(a) Use a letter or letters from the grid to identify:

(i) metabolic product(s) of respiration _____ **1**

(ii) hydrogen carrier(s) _____ **1**

(iii) source(s) of energy for the electron transport chain _____ **1**

(iv) material(s) derived from glucose _____ **1**

(b) Name the exact location of the electron transport chain in cells. **1**

6. Answer **either A or B**.

A Give an account of the structure and function of membranes in cells. **4**

OR

B Give an account of competitive and non-competitive inhibition of enzymes. **4**

Labelled diagrams may be used where appropriate.

MARKS | DO NOT WRITE IN THIS MARGIN

7. Rufous hummingbirds migrate thousands of kilometres each year between their summer breeding areas in Canada and their wintering areas in Mexico.

They feed on nectar throughout the year and save energy at night by entering a temporary state known as torpor in which body temperature and respiration rate are greatly reduced.

The chart below shows the average body mass of the hummingbirds and the number of hours per night spent in torpor throughout the year

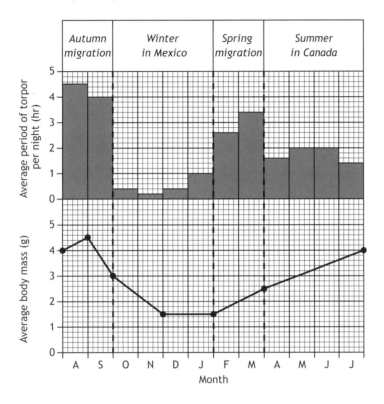

(a) (i) **Use values from the chart** to describe the changes in average body mass of the hummingbirds from the beginning of August until the end of January.

2

_____

_____

_____

MARKS DO NOT WRITE IN THIS MARGIN

7. **(a)** **(continued)**

(ii) Calculate the percentage increase in average body mass during the summer in Canada.

1

Space for calculation

_____ %

(iii) Suggest **one** reason for the increase in body mass of the birds during summer.

1

_____

_____

(b) (i) Suggest why the increased time spent in torpor during migration is an advantage to the birds.

1

_____

_____

(ii) Calculate the average period of torpor per month throughout the winter in Mexico.

1

Space for calculation

_____ hours per night

MARKS | DO NOT WRITE IN THIS MARGIN

7. (continued)

(c) The **table** below shows how the average oxygen consumption of the birds at rest is affected by their body temperature.

Body temperature at rest	Average oxygen consumption (cm³ per gram of body mass per hour)
Normal	15.0
Lowered during torpor	2.0

Using information from the **chart** and the **table,** calculate the volume of oxygen consumed per hour by a hummingbird, at the end of September, at normal body temperature.

1

Space for calculation

_____ cm³

(d) Describe a technique which could be used to track species such as the rufous hummingbird which undertake long distance migrations

1

_____

_____

8. The graph shows the growth phases in a culture of microorganisms in a previously sterilised fermenter.

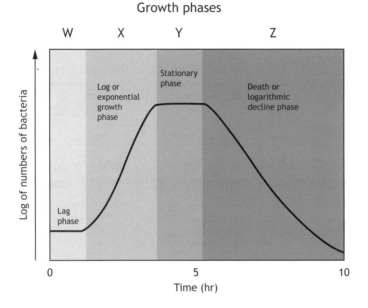

Growth phases

(a) (i) Using **letters** from the graph, complete the table to identify the phases in which the events shown occur.   **2**

Event	Letter
Secondary metabolites synthesised	
Induction of enzymes	

(ii) Describe the role of secondary metabolites and enzyme induction in the growth of microorganisms.

Secondary metabolites _____   **1**

_____

Enzyme induction _____   **1**

_____

MARKS | DO NOT WRITE IN THIS MARGIN

8. (continued)

(b) Explain why the numbers of microorganisms are plotted on a logarithmic scale.

1

_____

_____

(c) Apart from sterility, give two conditions which must be controlled to give optimum growth in a fermenter.

2

1 _____

2 _____

9. An experiment on the effect of light on the Calvin cycle was carried out using *Chlorella*, a unicellular alga.

500 cm^3 of *Chlorella* was placed in a glass flask in illuminated conditions for ten minutes and allowed to photosynthesise normally as shown in the diagram below.

Lamp

Flask with a suspension of the alga *Chlorella* in water

Tap

Collecting vessel

After this time, samples were withdrawn from the flask every minute for 5 minutes.

The lamp was then switched off and a further set of samples taken every minute for 5 minutes.

The concentration of RuBP and G3P in each sample was determined.

MARKS | DO NOT WRITE IN THIS MARGIN

9. **(continued)**

The results are shown in the table below.

Time (mins)	Illumination	RuBP concentration (units)	G3P concentration (units)
1	Yes	8	20
2	Yes	8	20
3	Yes	8	20
4	Yes	8	20
5	Yes	8	20
6	No	11	18
7	No	14	16
8	No	17	14
9	No	20	12
10	No	20	12

(a) (i) Identify the independent variable in this investigation.   1

_____

(ii) Identify **two** variables which should have been kept constant within the flask during the experiment to ensure that the procedure was valid.   1

1 _____

2 _____

(b) (i) Give a reason why the flask was left for 10 minutes before the sampling started.   1

_____

(ii) Suggest an advantage of using algae in an experiment of this type.   1

_____

_____

MARKS | DO NOT WRITE IN THIS MARGIN

9. (continued)

(c) Describe how the experimental procedure could be improved to increase the reliability of the results.

1

_____

_____

(d) On the grid below, complete the line graph to show the concentration of RuBP against time.

2

(Additional graph paper, if required, will be found on *Page twenty-one*)

(e) Using the results from the table, describe the evidence which suggests that RuBP is converted to G3P during the Calvin cycle of photosynthesis.

2

_____

_____

MARKS | DO NOT WRITE IN THIS MARGIN

10. Chimpanzees *Pan troglodytes* are primates which live in social groups with well developed social hierarchy. The table below lists facial expressions used in individual social interactions within the group.

Facial expression		Interaction in which expression is used
1		begging another individual for food
2		displaying fear or excitement
3		approaching a more dominant chimpanzee

(a) Give the term used to describe a list of this type.   1

_____

(b) Using information in the table, predict how the facial expression of a chimpanzee would change as it approached a more dominant individual then started to beg for food.   2

_____

_____

(c)   (i) Give **one** advantage to chimpanzees of living in social groups.   1

_____

_____

MARKS | DO NOT WRITE IN THIS MARGIN

10. (c) (continued)

(ii) Explain why long periods of parental care are needed for chimpanzee development.

1

(iii) Explain the advantage to chimpanzees of complex social behaviours such as the use of facial expressions.

1

11. The signal crayfish, *Pacifastacus leniusculus*, is a North American species. It was introduced to Europe in the 1970s to supplement native crayfish fisheries. It has spread rapidly in its new habitats and is now considered an invasive species across Europe.

(a) Explain why the population of invasive species can become very large in their new habitats.

1

(b) Give **two** types of damage which large populations of invasive species can inflict on the biodiversity of an ecosystem in which they have become established.

1 _____   1

2 _____   1

(c) Plans are being made to control the signal crayfish population in Scotland.

Suggest a method of biological control by which the crayfish numbers might be reduced and give a possible drawback with this method.

Method _____   1

Possible drawback _____   1

MARKS | DO NOT WRITE IN THIS MARGIN

12. Pure bred cattle tend to be affected by inbreeding depression and have poorer health and vigour characteristics than crossbreeds.

The chart below shows the average milk yield from two breeds of cattle and their crossbreed offspring.

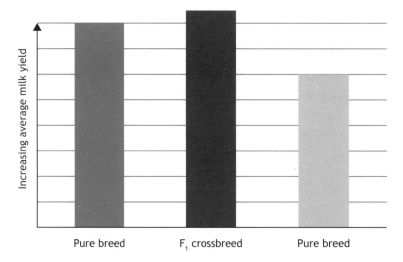

Increasing average milk yield

Pure breed   F₁ crossbreed   Pure breed

(a) Describe what the chart shows about the milk yield of the three breeds.   1

_____

_____

(b) Explain the following statements in terms of overall food security.

   (i) The F₁ cattle are not bred together to produce F₂ animals for milk.   2

   _____

   _____

   _____

   (ii) It is an advantage to grow crops on an area of fertile soil than to use the area as pasture for raising cattle.   1

   _____

   _____

MARKS | DO NOT WRITE IN THIS MARGIN

12. **(continued)**

(c) Describe why each of the following types of cross would be carried out as part of a cattle-breeding programme.

Test cross _____   1

_____

Back cross _____   1

_____

13. Long-tailed tits *Aegithalos caudatus* are small birds which breed in loose colonies in which many of the birds are related.

If a pair fails in breeding, they will often behave in an altruistic way by assisting neighbouring pairs to feed their chicks.

Give the meaning of the term "altruistic" and, from the information given, suggest an explanation for the birds' behaviour.

Meaning _____   1

_____

Explanation _____   2

_____

MARKS | DO NOT WRITE IN THIS MARGIN

14. Answer **either A or B.**

A   Give an account of the evolution of new species under the following headings:

   (i)   genetic drift;   3

   (ii) natural selection.   6

OR

B   Give an account of DNA molecules under the following headings:

   (i)   their organisation in living cells;   4

   (ii) their amplification by the polymerase chain reaction (PCR).   5

Labelled diagrams may be used where appropriate.

[END OF MODEL PAPER]

ADDITIONAL GRAPH PAPER FOR QUESTION 9 (d)

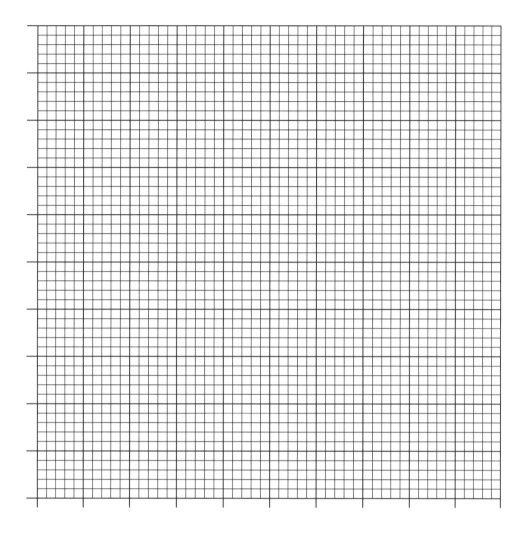

ADDITIONAL SPACE FOR ANSWERS AND ROUGH WORK

ADDITIONAL SPACE FOR ANSWERS AND ROUGH WORK

ADDITIONAL SPACE FOR ANSWERS AND ROUGH WORK

HIGHER FOR CfE

2015

National
Qualifications
2015

X707/76/02

Biology
Section 1—Questions

WEDNESDAY, 13 MAY

1:00 PM – 3:30 PM

Instructions for the completion of Section 1 are given on *Page two* of your question and answer booklet X707/76/01.

Record your answers on the answer grid on *Page three* of your question and answer booklet.

Before leaving the examination room you must give your question and answer booklet to the Invigilator; if you do not, you may lose all the marks for this paper.

SECTION 1 — 20 marks
Attempt ALL questions

1. Which line in the table below shows features of the human genome?

	Contains base sequences that regulate transcription	Contains base sequences transcribed to RNA but never translated	Contains base sequences from which primary transcripts are produced
A	✗	✓	✗
B	✗	✗	✓
C	✓	✓	✗
D	✓	✓	✓

2. The diagram below shows a eukaryotic gene containing introns and exons and a scale bar representing the number of base pairs in the gene.

Number of base pairs

How many bases will there be in the mature mRNA formed from the primary transcript of this gene?

A 180

B 540

C 560

D 720

3. Which of the following would **not** explain loss of genetic diversity in a population?

A Inbreeding

B The founder effect

C The bottleneck effect

D No barriers to gene flow

4. The following are events in the evolution of life on Earth.

   1   Animals appear
   2   Vertebrates appear
   3   Land plants appear

   In which order are these events thought to have occurred?

   A   1 2 3

   B   1 3 2

   C   3 1 2

   D   3 2 1

5. The graph below shows a molecular clock which compares the amino acid sequences in the protein cytochrome C in various vertebrate groups.

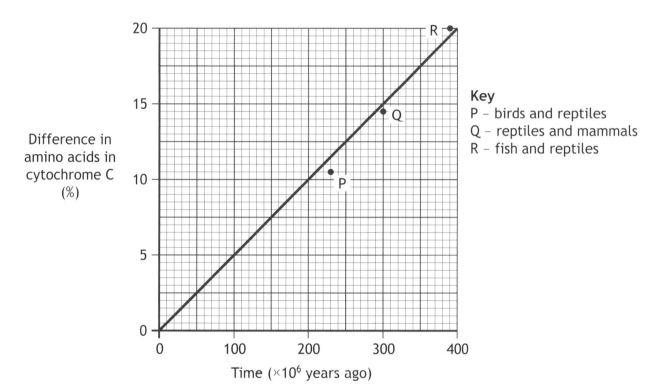

Key
P – birds and reptiles
Q – reptiles and mammals
R – fish and reptiles

Difference in amino acids in cytochrome C (%)

Time (×10⁶ years ago)

   From the information in the graph, which vertebrate groups shared a common ancestor most recently?

   A   Fish and reptiles

   B   Birds and mammals

   C   Reptiles and mammals

   D   Birds and reptiles

[Turn over

6. The melting temperature of a molecule of DNA (T_m) is the temperature at which half of its base pairs separate. T_m is proportional to the percentage of the guanine to cytosine (G–C) base pairs in the molecule as shown on the graph below.

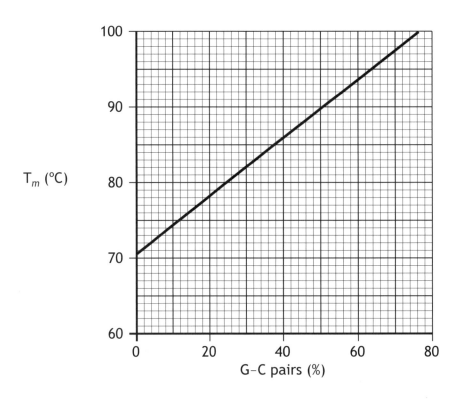

T_m (°C)

G–C pairs (%)

The numbers of base pairs present in a DNA molecule are shown in the table below.

Number of base pairs present	
A–T	G–C
1200	800

What is T_m for this molecule?

A   78 °C

B   86 °C

C   94 °C

D   96 °C

7. The following are molecules that can be broken down into substrates for respiration.

   1   starch
   2   protein
   3   fat

   Which molecules can be broken down into products which can be converted directly into intermediates of the citric acid cycle?

   A   1 only

   B   1 and 3 only

   C   2 and 3 only

   D   1, 2 and 3

8. The effect of an antibiotic on a bacterial species was tested by spreading a culture of each of the bacterial species on agar plates and adding a disc of absorbent paper soaked in the antibiotic, as shown in the diagram below.

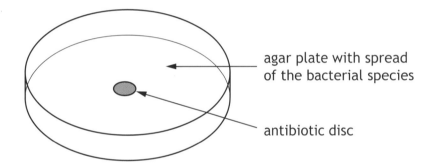

   The plate was incubated for 24 hours at 30 °C and the growth examined.

   Which of the following would be a suitable control for this experiment?

   Repeat the experiment exactly but

   A   with no bacteria

   B   incubate at human body temperature

   C   use a disc with no antibiotic

   D   use a disc with a different antibiotic.

9. Mitochondria are small membrane-bound compartments present in eukaryotic cells.

   One advantage to a mammalian muscle cell of having many small mitochondria is that they provide a

   A   small surface area to volume ratio to increase the uptake of oxygen

   B   large surface area to volume ratio to increase the uptake of oxygen

   C   large surface area to volume ratio to decrease the uptake of carbon dioxide

   D   small surface area to volume ratio to decrease the uptake of carbon dioxide.

10. When salmon migrate from freshwater into seawater, changes in concentration of their surroundings are detected and the activity of the ion pumps in the salmon gills increases. The activity of the ion pumps decrease when the salmon migrate back to freshwater.

Which line in the table below shows the description of the salmon and the control of its ion pumps?

	Description of salmon	Control of ion pumps
A	conformer	by negative feedback
B	conformer	behavioural
C	regulator	by negative feedback
D	regulator	behavioural

11. The rate of sweat production of two individuals, X and Y, was measured during and after a period of exercise.

   The results are shown in the graph below.

   Which of the following conclusions can be drawn from the graph?

   A   The rate of sweat production of individual X is always greater than individual Y.

   B   Individuals X and Y both reach their maximum sweat production at 20 minutes.

   C   Individual X starts increasing sweat production sooner than individual Y.

   D   The greatest difference in sweat production by individuals X and Y is at 50 minutes.

[Turn over

12. The table below shows the results of pharmacogenetic tests on a drug designed to treat a liver infection in a group of patients.

		Number of patients	
		beneficial effect on patient	no beneficial effect on patient
Number of patients	toxic side-effects	30	15
	no side-effects	60	45

What percentage of the patients gained benefit from the drug but showed toxic side-effects?

A   20

B   25

C   30

D   90

13. The statements below give information on three different bacterial species.

1   *Psychrobacter adeliensis* is found in Antarctica. It has been isolated from coastal ice and grows well at low temperatures.

2   *Thermophilus aquaticus* lives in hot springs and generates ATP by removal of high energy electrons from inorganic molecules.

3   *Escherichia coli* has enzymes with an optimal temperature of 37 °C. Most strains of this species are harmless and live in animal intestines although some strains can be harmful to the host animal.

From this information, which of these bacterial species can be classified as extremophile?

A   1 and 2 only

B   1 and 3 only

C   2 only

D   3 only

14. Which of the following results in a transfer of electrons down the electron transport chains during the light dependent reactions of photosynthesis?

A   NADP is converted to NADPH

B   Water is split by photolysis

C   ATP is synthesised

D   Pigment molecules absorb energy

15. When quantifying plant productivity, the economic yield is the

A   total biomass produced

B   biomass of desired product

C   increase in biomass due to photosynthesis

D   rate of biomass production per hectare.

16. Soil type is dependent on the composition of its components which in turn affects the productivity of plants growing in it.

The table below shows the percentage of each component present in four different soils.

Soil type	Component (%)		
	clay	silt	sand
sandy clay loam	20–30	0–30	50–80
clay loam	20–35	20–60	20–50
sandy silt loam	0–20	40–80	20–50
silty clay loam	20–35	45–80	0–20

Which of the following charts represents a clay loam?

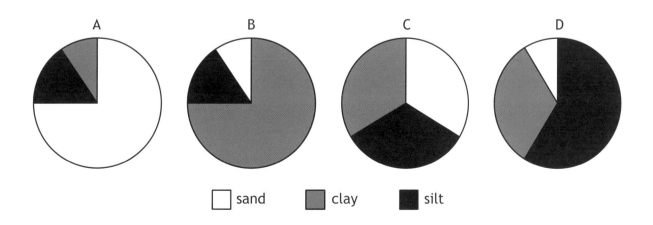

sand   clay   silt

17. The table below shows the number of beet armyworm larvae found in plots of cotton plants.

Some plots were treated with insecticide on 27 June and 1 August and other plots left untreated.

Sampling date		Number of beet armyworm larvae	
		Treated plots	Untreated plots
July	8	3	3
	15	33	2
	22	22	17
	29	42	10
August	5	120	8
	12	160	10

Which of the following is the most likely explanation for the differences between the treated and untreated plots?

A   The insecticide kills a predator of the larvae

B   The larvae are resistant to the insecticide

C   The beet armyworm breeds in July

D   The larvae have a short lifecycle

18. In primates such as chimpanzees, parental care

A   occurs over a short time period

B   provides time for learning complex social behaviour

C   increases the parent's social status within their group

D   involves appeasement behaviour within a group.

19. Altruistic behaviour between closely related animals

A   reduces competition between individuals in the population

B   increases the survival chances of the donor animal

C   increases the frequency of shared genes in the next generation

D   reduces unnecessary aggression and conflict in social groups.

20. A species that plays a role vital for the survival of many other species in an ecosystem is called

A   a keystone species

B   a native species

C   an invasive species

D   a dominant species.

**[END OF SECTION 1. NOW ATTEMPT THE QUESTIONS IN SECTION 2
OF YOUR QUESTION AND ANSWER BOOKLET]**

[BLANK PAGE]

DO NOT WRITE ON THIS PAGE

H
National Qualifications 2015

Mark

X707/76/01

Biology
Section 1 — Answer Grid and Section 2

WEDNESDAY, 13 MAY
1:00 PM – 3:30 PM

Fill in these boxes and read what is printed below.

Full name of centre

Town

Forename(s)

Surname

Number of seat

Date of birth

Day	Month	Year	Scottish candidate number

Total marks — 100

SECTION 1 — 20 marks

Attempt ALL questions.

Instructions for completion of Section 1 are given on *Page two*.

SECTION 2 — 80 marks

Attempt ALL questions.

Questions 5 and 13 each contain a choice.

Write your answers clearly in the spaces provided in this booklet. Additional space for answers and rough work is provided at the end of this booklet. If you use this space you must clearly identify the question number you are attempting. Any rough work must be written in this booklet. You should score through your rough work when you have written your final copy.

Use **blue** or **black** ink.

Before leaving the examination room you must give this booklet to the Invigilator; if you do not you may lose all the marks for this paper.

SECTION 1— 20 marks

The questions for Section 1 are contained in the question paper X707/76/02.
Read these and record your answers on the answer grid on *Page three* opposite.
Use **blue** or **black** ink. Do NOT use gel pens or pencil.

1.   The answer to each question is **either** A, B, C or D. Decide what your answer is, then fill in the appropriate bubble (see sample question below).

2.   There is **only one correct** answer to each question.

3.   Any rough working should be done on the additional space for answers and rough work at the end of this booklet.

Sample Question

The thigh bone is called the

     A   humerus

     B   femur

     C   tibia

     D   fibula.

The correct answer is **B**—femur. The answer **B** bubble has been clearly filled in (see below).

Changing an answer

If you decide to change your answer, cancel your first answer by putting a cross through it (see below) and fill in the answer you want. The answer below has been changed to **D**.

If you then decide to change back to an answer you have already scored out, put a tick (✓) to the **right** of the answer you want, as shown below:

SECTION 1 — Answer Grid

	A	B	C	D
1	○	○	○	○
2	○	○	○	○
3	○	○	○	○
4	○	○	○	○
5	○	○	○	○
6	○	○	○	○
7	○	○	○	○
8	○	○	○	○
9	○	○	○	○
10	○	○	○	○
11	○	○	○	○
12	○	○	○	○
13	○	○	○	○
14	○	○	○	○
15	○	○	○	○
16	○	○	○	○
17	○	○	○	○
18	○	○	○	○
19	○	○	○	○
20	○	○	○	○

[BLANK PAGE]

DO NOT WRITE ON THIS PAGE

[Turn over for Question 1 on *Page six*

DO NOT WRITE ON THIS PAGE

MARKS | DO NOT WRITE IN THIS MARGIN

SECTION 2 — 80 marks

Attempt ALL questions

Questions 5 and 13 each contain a choice.

1. Sauerkraut is a food produced by preserving cabbage. Preservation involves inhibition of the bacteria which can spoil the food. *Lactobacillus* is anaerobic and, unlike most bacteria, grows well at low pH.

   The diagram below shows stages in fermentation of the glucose in cabbage by *Lactobacillus*.

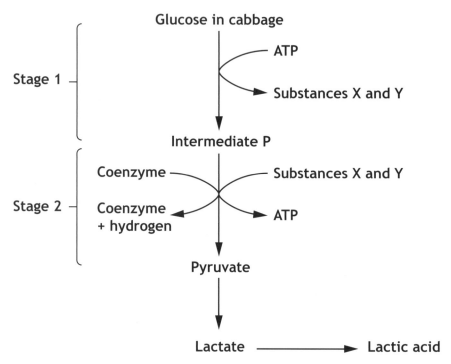

   (a) (i) Identify substances **X** and **Y** in the diagram.   1

   X _____

   Y _____

   (ii) Name the coenzyme, shown in the diagram, which carries hydrogen to the electron transport chain.   1

   _____

   (iii) Explain why the ATP produced at **Stage 2** in the diagram is referred to as an energy pay off.   1

   _____

   _____

MARKS | DO NOT WRITE IN THIS MARGIN

1. (continued)

(b) The flow chart below shows how cabbage can be processed to produce sauerkraut.

Step 1
| Cabbage is shredded and packed in airtight bags |

Step 2
| *Lactobacillus* present on the cabbage produces lactic acid which lowers the pH inside the bags |

Step 3
| Cabbage is cooked and pasteurised inside the bags to produce sauerkraut |

(i) Explain why shredding the cabbage in **Step 1** increases the rate of production of sauerkraut.

1

_____

_____

(ii) Explain why the process encourages the growth of *Lactobacillus* but inhibits the growth of other bacteria.

1

_____

_____

[Turn over

Page seven

MARKS | DO NOT WRITE IN THIS MARGIN

2. Erythropoietin (EPO) is a protein synthesised in the kidneys which is involved in red blood cell production. Some individuals with kidney disease have low red blood cell counts and can be treated with EPO.

EPO can be produced by recombinant DNA technology in which the human EPO gene was inserted into a specially prepared bacterial plasmid.

The diagram below shows the prepared bacterial plasmid before and after it was modified by the insertion of a human EPO gene.

Prepared plasmid before modification   **Modified plasmid**

(a) Explain the importance of removing the EPO gene from a human chromosome with the **same** restriction endonuclease that was used to open the bacterial plasmid.

1

_____

_____

(b) Name the enzyme used to seal the EPO gene into the bacterial plasmid.

1

_____

(c) Modified plasmids were mixed with bacteria. Some bacterial cells were transformed by taking up the modified plasmids but others were not.

Use information from the diagram to suggest how a culture containing only the transformed bacteria was obtained.

1

_____

_____

(d) Identify the section of the modified plasmid shown in the diagram which ensured that it could be copied and passed to daughter cells when transformed bacteria divided.

1

_____

MARKS

2. **(continued)**

(e) The EPO protein produced by the transformed bacteria is inactive.

(i) Suggest a reason why bacteria produce EPO protein which is inactive.

1

_____

_____

(ii) Suggest how recombinant DNA technology could be used to produce an active form of the EPO protein.

1

_____

_____

[Turn over

3. (a) The yeast *Kluyveromyces marxianus* uses lactose as a respiratory substrate. An investigation was carried out into the effect of lactose concentration on ethanol production by this yeast species. Five flasks were set up each containing $5\,cm^3$ of yeast suspension and $100\,cm^3$ of 4, 8, 12, 16 or 20% lactose solution. The flasks were sealed to maintain anaerobic conditions.

Samples were removed from each flask at 12 and 36 hours and the concentration of ethanol was determined. Results are shown in the table below.

Lactose concentration (%)	Ethanol concentration (g per $100\,cm^3$)	
	12 hours	36 hours
4	1·20	1.65
8	1·55	2·80
12	2·00	4·25
16	2·80	3·25
20	2·80	6·50

(i) Identify the independent variable.   1

_____

(ii) Identify **one** variable not already mentioned that should be kept constant so that a valid conclusion can be drawn.   1

_____

(b) Describe the relationship between the lactose concentration and ethanol concentration at 12 hours growth.   2

_____

_____

(c) Calculate the percentage increase in ethanol concentration between 12 and 36 hours growth in the 4% lactose flask.   1

Space for calculation

_____ %

MARKS | DO NOT WRITE IN THIS MARGIN

3. **(continued)**

(d) Air leaked into the 16% lactose flask between 12 and 36 hours growth. Explain why this resulted than a lower than expected ethanol concentration.

1

_____

_____

_____

[Turn over

MARKS | DO NOT WRITE IN THIS MARGIN

4. The northern blossom bat *Macroglossus minimus* is an Asian species which has a high metabolic rate and a daily rhythm of torpor.

The metabolic rates and body temperatures of a group of these bats were recorded every four hours over a 24 hour cycle and the results are shown on the graph below.

(a) Calculate the oxygen consumption of a 16 g bat at 00:00 hours.

Space for calculation

1

_____ cm³ O₂ per hr

(b) Tick (✓) **one** box to identify the period when the bats were in full torpor and justify your answer.

2

16:00 – 20:00   20:00 – 00:00   04:00 – 08:00   08:00 – 12:00

☐   ☐   ☐   ☐

Justification_____

_____

MARKS | DO NOT WRITE IN THIS MARGIN

4.   (continued)

(c)   Give **one** benefit to bats of their daily torpor.

1

(d)   Blossom bats are nocturnal.

Give **one other** behavioural adaptation of animals with high metabolic rates to allow survival in adverse conditions.

1

[Turn over

MARKS | DO NOT WRITE IN THIS MARGIN

5. Answer **either A or B** in the space below.

A   Describe the arrangement of heart chambers in birds and amphibians and relate this to their metabolic rates.   **4**

OR

B   Describe competitive and non-competitive inhibition of enzyme action.   **4**

[Turn over for Question 6 on *Page sixteen*

DO NOT WRITE ON THIS PAGE

MARKS | DO NOT WRITE IN THIS MARGIN

6. The diagram below shows some stages in the Calvin cycle of photosynthesis.

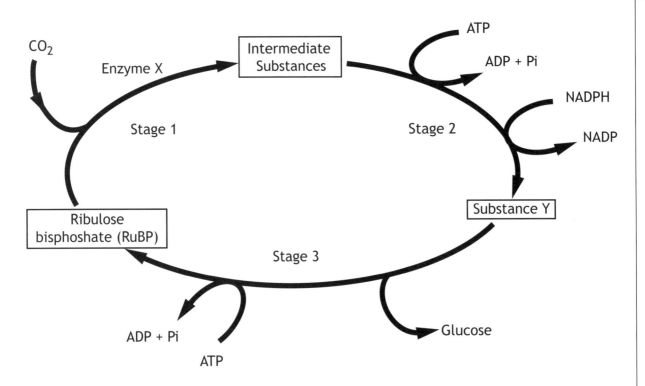

(a) (i) Name Enzyme X and Substance Y.   2

Enzyme X _____

Substance Y_____

(ii) Explain the importance of producing glucose and RuBP in Stage 3.   2

Glucose_____

_____

RuBP_____

(b) Research has been carried out which aims to increase photosynthesis in crop plants by inserting genes for the production of prokaryotic pigments into the cells. These pigments absorb wavelengths of light which are different to those absorbed by the pigments present in the crop plants.

(i) Predict what would happen to the concentrations of ATP and NADPH in the crop plant cells.   2

ATP   _____

NADPH   _____

MARKS | DO NOT WRITE IN THIS MARGIN

6. **(b)** **(continued)**

(ii) Genetically modified (GM) crops are evaluated in field trials.

Certain experimental procedures are required when setting up field trials to compare GM and non GM crops.

Give **one** such procedure and explain how it allows valid conclusions to be drawn.

2

Procedure _____

_____

Explanation _____

_____

[Turn over

MARKS | DO NOT WRITE IN THIS MARGIN

7.   The parasite *Schistosoma mansoni* causes the condition schistosomiasis in humans.

The condition is common in tropical regions where the parasite is often present in fresh water.   Humans can be infected if they enter water containing the parasite.

The life cycle of *Schistosoma mansoni* is shown below.

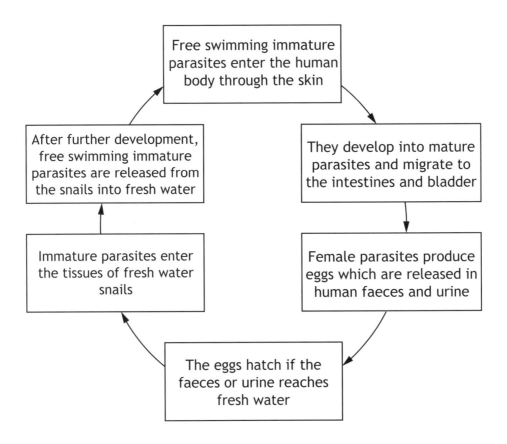

(a)   Explain why *Schistosoma mansoni* is described as a parasite.   1

_____

(b)   Identify the secondary host and suggest a benefit to *Schistosoma mansoni* of including a secondary host in its life cycle.   2

Secondary host _____

Benefit _____

_____

(c)   Describe **one** measure which could be adopted to reduce the number of cases of schistosomiasis.   1

_____

MARKS | DO NOT WRITE IN THIS MARGIN

8. Harlequin ladybirds, *Harmonia axyridis*, were introduced to the UK from their native habitat in Eastern Asia in order to reduce the population of aphids, which feed on crop plants.

Since their introduction, harlequin ladybirds have spread rapidly and their population has dramatically increased. As a result the populations of some ladybird species have dramatically decreased, although the population of native seven-spot ladybirds has remained relatively stable.

(a) Name this control method used to manage the population of aphids. 　1

(b) Using the information given, explain why the harlequin ladybird can be described as an invasive species. 　1

(c) Suggest one reason why the population size of the seven-spot ladybird has remained relatively stable. 　1

(d) Give a reason why the population of harlequin ladybirds has increased more quickly in the UK than in their native habitat. 　1

[Turn over

MARKS | DO NOT WRITE IN THIS MARGIN

9. Alfalfa is a crop plant often grown for cattle food.

In a field trial, alfalfa was grown in six plots each of which had been treated with a different level of phosphate fertiliser. The alfalfa was harvested after 24 weeks of growth and the total dry mass of the crop at each fertiliser level was calculated. The protein content of the alfalfa grown at each fertiliser level was determined.

The results are shown in the graph below.

(a) (i) **Use values from the graph** to describe the changes in the yield of total dry mass of the crop as the phosphate fertiliser level was increased from 0 to 100 kg per hectare.

2

_____

_____

_____

_____

(ii) Predict the protein content of an alfalfa crop if 120 kg of phosphate fertiliser per hectare had been applied.

1

_____ kg per tonne dry mass

9. (a) (continued)

(iii) Calculate the total mass of protein produced from one hectare when 40 kg of phosphate fertiliser per hectare was applied.

Space for calculation

_____ kg

(b) In a feeding trial, three groups of 10 cattle were fed with alfalfa of different protein contents over a 25 day period. The cattle were weighed at the beginning and end of this period and the average increase in their body mass calculated.

The results are shown in the table below.

Cattle group	Protein content of alfalfa fed to cattle (kg per tonne dry mass)	Average increase in body mass of cattle over a 25 day period (kg)
1	80	12
2	90	15
3	120	17

(i) State how the design of the feeding trial ensured the reliability of the results.

_____

(ii) Using the information from the **table**, calculate the average increase in body mass per day of the cattle in Group 2.

Space for calculation

_____ kg per day

(iii) Using information from the **graph and table**;

1 suggest the phosphate fertiliser level which was applied in the production of the alfalfa which the cattle in Group 2 were fed;

_____ kg per hectare

2 draw a conclusion about how phosphate fertiliser levels applied to the alfalfa affected the growth of cattle in the feeding trial.

_____

MARKS

1

1

1

1

1

MARKS | DO NOT WRITE IN THIS MARGIN

9. (continued)

(c) In terms of food security, explain why using agricultural land to grow cereal for human consumption rather than to grow cattle food would produce more food for humans per unit area.

1

_____

_____

[Turn over for Question 10 on *Page twenty-four*

DO NOT WRITE ON THIS PAGE

MARKS | DO NOT WRITE IN THIS MARGIN

10. *Staphylococcus aureus* (*S.aureus*) is a species of bacteria that lives on human skin. This species of bacteria can cause infections if it enters the body through a wound. *S.aureus* infections can be treated with antibiotics such as methicillin and penicillin.

Infections can be caused by a strain of *S.aureus* called MRSA which is resistant to methicillin and penicillin and is becoming more common.

(a) The MRSA strain has developed resistance to antibiotics by gene transfer from another organism.

Identify the correct statement(s) relating to MRSA antibiotic resistance.

Tick (✓) the correct box(es).   **2**

MRSA has developed antibiotic resistance through **horizontal** gene transfer from another organism.	
MRSA has developed antibiotic resistance through **vertical** gene transfer from another organism.	
This type of gene transfer in bacteria brings about a **rapid** evolutionary change.	
This type of gene transfer in bacteria brings about a **slow** evolutionary change.	

(b) Explain how the overuse of antibiotics has led to the increase in the population of MRSA.   **2**

_____

_____

_____

MARKS | DO NOT WRITE IN THIS MARGIN

10. **(continued)**

(c) Samples were taken from a patient suspected of having a bacterial infection. The samples were used to inoculate plates of agar as shown in the diagram below.

Nutrient agar only

Nutrient agar + penicillin

Nutrient agar + methicillin

(i) Predict the results if the cause of the bacterial infection was MRSA.

1

_____

(ii) The nutrient agar contained specific amino acids required for protein synthesis. Suggest **one other** type of complex compound that the nutrient agar may have contained.

1

_____

[Turn over

MARKS | DO NOT WRITE IN THIS MARGIN

11. Patients requiring an organ transplant are tissue typed to match with potential donors. Polymerase chain reaction (PCR) and gel electrophoresis are used to compare DNA sequences of the patient with those of donors. Gel electrophoresis separates mixtures of DNA fragments according to size. The presence of a specific DNA band indicates that a donor is a suitable match.

Patient and potential donor samples were compared with a DNA ladder.

The DNA ladder contains fragments of DNA, separated by gel electrophoresis, which are of a known size and measured in base pairs (bp). The distances the DNA fragments travelled were measured and are shown in the table below. The diagram below shows the result of the gel electrophoresis.

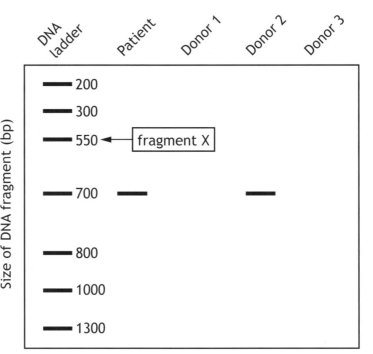

Size of DNA fragment (bp)	Distance travelled (mm)
200	72
300	58
550	32
700	18
800	12
1000	10
1300	8

(a) The gel used for electrophoresis contains agarose. Calculate the mass of agarose required to make $30\,cm^3$ of a 0·8% agarose gel.

Space for calculation

1

_____ g

(b) Using information in the **table** and the **diagram** give the distance travelled by fragment X in the DNA ladder.

1

_____ mm

11. (continued)

(c) On the grid below, draw a line graph to show the distance travelled against the size of DNA fragment.   **2**

(Additional graph paper if required will be found on *Page thirty-three*.)

(d) Give a conclusion about the suitability of the donors.   **1**

_____

(e) (i) The base sequence of a primer used in the PCR procedure is shown below.

A T G A C A A A T C G

Give the base sequence of a DNA fragment to which this primer would bind.   **1**

_____

(ii) Complete the table below to show the temperatures used in two stages of the PCR procedure and the reasons for using these temperatures.   **2**

Temperature (°C)	Reason
94	
	Allows primer to bind to target sequence

MARKS | DO NOT WRITE IN THIS MARGIN

12. An investigation was carried out involving a number of patients with heart disease. A group of volunteer patients was treated with adult stem cells and a control group was not given this treatment.

Six weeks after the treatment, the average heart rate and the average volume of blood pumped out per heartbeat (stroke volume) was determined for each group.

The results are shown in the table below.

	Patients given stem cell treatment	Patients not given stem cell treatment
Average heart rate (beats per minute)	70	70
Average stroke volume (cm³)	45	28

(a) Give **two** conclusions which can be drawn about the effect of the stem cell treatment on the patients.   2

1 _____

_____

2 _____

_____

(b) Another important measure of heart performance is cardiac output.

$$Cardiac\ output = heart\ rate \times stroke\ volume$$
(cm³ per minute)   (bpm)   (cm³)

Calculate the average increase in cardiac output in those patients given the stem cell treatment compared to those in the control group.   1

Space for calculation

_____ cm³ per minute

MARKS

DO NOT WRITE IN THIS MARGIN

12. (continued)

(c) (i) Describe how tissue (adult) stem cells differ from embryonic stem cells.

1

_____

_____

(ii) Describe how the heart cells produced by the patients as a result of the stem cell treatment in this investigation developed their specialised functions.

1

_____

_____

(d) Much stem cell research is related to the therapeutic value of stem cells.

Give **one other** reason for carrying out stem cell research.

1

_____

_____

[Turn over for Question 13 on *Page thirty*

13. Answer **either A or B** in the space below and on pages *thirty-one* and *thirty-two*.

Labelled diagrams may be used where appropriate.

A   Write notes on gene expression in eukaryotes under the following headings:

    (i)   production of mRNA;   5

    (ii) translation of mRNA.   4

OR

B   Write notes on mutation under the following headings:

    (i)   single gene mutations;   4

    (ii) chromosome mutations and polyploidy.   5

SPACE FOR ANSWERS

SPACE FOR ANSWERS

[END OF QUESTION PAPER]

ADDITIONAL SPACE FOR ANSWERS AND ROUGH WORK

ADDITIONAL GRAPH PAPER FOR QUESTION 11(c)

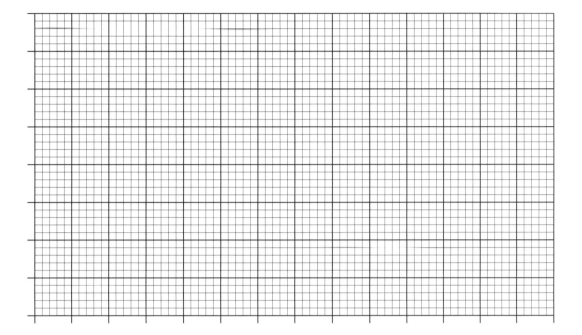

ADDITIONAL SPACE FOR ANSWERS AND ROUGH WORK

HIGHER FOR CfE | ANSWER SECTION

SQA AND HODDER GIBSON HIGHER FOR CfE
BIOLOGY 2015

HIGHER FOR CfE BIOLOGY
SPECIMEN QUESTION PAPER

Section 1

Question	Response	Mark
1.	C	1
2.	B	1
3.	C	1
4.	A	1
5.	D	1
6.	A	1
7.	A	1
8.	C	1
9.	B	1
10.	D	1
11.	A	1
12.	A	1
13.	B	1
14.	A	1
15.	D	1
16.	C	1
17.	D	1
18.	D	1
19.	A	1
20.	B	1

Section 2

Question			Expected response	Max mark
1.	(a)	(i)	• Intron/Intron 1/Intron 2	1
		(ii)	• (Alternative) RNA splicing	1
		(iii)	• Depending on which RNA segments are treated as exons and introns 1 different segments can be spliced together to produce different mRNA transcripts 1 OR • Appropriate example from diagram	2
	(b)		• Cutting and combining different protein chains 1 OR • Adding phosphate to the protein OR • Adding carbohydrate to the protein	1
2.	(a)		• Translocation	1
	(b)	(i)	• Competitive	1
		(ii)	• 95	1

Question			Expected response	Max mark
		(iii)	• Drug was effective as white blood count reduced to normal • Drug works by inhibiting the enzyme produced by Philadelphia chromosome	2
3.	(a)		• Stage 1 separates strands or breaks H bonds • Stage 2 allows primer to bond/anneal to strand/target sequence	2
	(b)		• 7	1
	(c)		• Identical set up but without primers	1
	(d)		• Forensic use/paternity testing	1
4.	(a)		• Sequence data	1
	(b)		• Horizontal/lateral	1
	(c)	(i)	• 25	1
		(ii)	• Last common ancestor of humans and chimpanzees was more recent than humans and orangutans • Chimpanzees and humans 5 million years ago, orangutans and humans 19 million years ago	2
5.	(a)		• P is Acetyl CoA • Q is Oxaloacetate	2
	(b)		• ATP/Energy is required 1 • A greater amount of energy/ATP is produced 1	2
	(c)		• Carry hydrogen and high energy electrons 1 • To the electron transport chain 1	2
	(d)		• Less ATP/energy is produced 1 • Fewer electrons are passed to electron transport chain OR • Fewer hydrogen ions are pumped through the membrane OR • ATP synthase is damaged 1	2
6.	(a)		• 20	1
	(b)		• Increase - people becoming complacent about hand washing or bacteria becoming resistant OR • No change - everyone now using procedure OR • Decrease - increased uptake of procedure	1
	(c)		• *Clostridium* increases • *Staphylococcus* remains fairly constant	2
	(d)		• Conclusion - effective • Justification - although percentage of cases remains similar number of cases falls	2

Question			Expected response	Max mark
	(e)		• Type - *Clostridium* • Reason - percentage of cases due to *Clostridium* increased	1
7.	(a)		• Enzymes have an optimum temperature or only work within a certain temperature range	1
	(b)	(i)	• Hypothalamus	1
		(ii)	• Nerve (impulse)	1
	(c)	(i)	• Vasoconstriction/vessels get narrower	1
		(ii)	• Reduces blood flow to skin so less heat loss	1
8.	A		• 1 metabolic rate reduced • 2 dormancy can be predictive or consequential • 3 hibernation in winter usually mammals • 4 aestivation allows survival in periods of drought or high temperature • 5 daily torpor is reduced activity in animals with high metabolic rates • 6 example of hibernation or aestivation or daily torpor	4
	B		• 1 plant/animal gene transferred into microorganism that makes plant/animal protein • 2 restriction endonuclease to cut gene out/cut plasmid • 3 genes introduced to increase yield or prevent microbe surviving in external environment • 4 ligase seals gene into plasmid • 5 recombinant yeast cells to overcome polypeptides being incorrectly folded or lacking post translational modifications • 6 regulatory sequences in plasmids/artificial chromosomes to control gene expression	4
9.	(a)		• 15	1
	(b)		• 413·44	1
	(c)		• Milk yield/fat content increased by crossbreeding • Protein content decreased by crossbreeding	1
	(d)		• Inbreeding depression	1
	(e)	(i)	• F2 has a variety of genotypes	1
		(ii)	• Selection or backcrossing	1
10.	(a)		• Rate of photosynthesis	1
	(b)		• Use a water bath	1
	(c)		• Easier to separate algae from solution OR • Easier to control algae concentration	1
	(d)		• Repeat at each distance/light intensity	1
	(e)		• Axes and labels • Plotting and joined with a ruler	2

Question			Expected response	Max mark
	(f)		• As light intensity increases rate increases • At higher light intensities rate remains constant	2
11.	(a)		• Worker bees are related to queen's offspring • So worker bees share genes with queen's offspring	2
	(b)	(i)	• Increase from 4.2 million (in 1980) to 4.4 million (in 1985) then decrease to 2.8 million (in 1995)	1
		(ii)	• 2:1	1
12.	(a)	(i)	• Number/frequency of alleles in a population	1
		(ii)	• Small population may lose the genetic variation necessary to enable evolutionary responses to environmental change OR • The loss of genetic diversity can lead to inbreeding which results in poor reproductive rates	1
	(b)		• Edge species may invase the interior of the habitat and compete with interior species	1
	(c)	(i)	• Area of natural habitat linking fragments	1
		(ii)	• Individual members of the locally extinct species can move into the fragment and recolonise	1
13.	(a)		• Invasive	1
	(b)		• Light OR • Water OR • Minerals OR • Nutrients	1
	(c)	(i)	• May eat native plants OR • May become invasive	1
		(ii)	• Test effect on native species in an enclosed area	1

Question			Expected response	Max mark
14.	A		• 1 double strand of nucleotides/ double helix • 2 deoxyribose sugar, phosphate and base • 3 sugar phosphate backbone • 4 complementary bases pair or A-T and C-G • 5 H bonds between bases • 6 antiparallel structure with deoxyribose and phosphate at 3' and 5' ends • 7 proteins/histones • 8 DNA unwinds into 2 strands • 9 primer needed to start replication • 10 DNA polymerase adds nucleotides to 3' (deoxyribose) end of strand • 11 DNA polymerase adds nucleotides in one direction • 12 one strand replicated continuously, the other in fragments • 13 fragments joined by ligase	9
	B		• 1 isolation barriers prevent gene flow between populations/populations interbreeding • 2 geographical isolation leads to allopatric speciation • 3 behavioural isolation leads to sympatric speciation • 4 ecological isolation leads to sympatric speciation • 5 different mutations occur on each side of isolation barrier • 6 some mutations may be favourable • 7 natural selection is non-random increase in frequency of genetic sequences that increase survival • 8 sexual selection is non-random increase in frequency of genetic sequences that increase reproductive success • 9 Any 2 from disruptional/ directional/stabilising selection • 10 third type of selection from 9 • 11 after many generations/long period of time • 12 new species form • 13 if populations can no longer interbreed to produce fertile young then different species	9

HIGHER FOR CfE BIOLOGY MODEL PAPER 1

Section 1

Question	Response	Mark
1.	D	1
2.	C	1
3.	A	1
4.	D	1
5.	A	1
6.	C	1
7.	A	1
8.	C	1
9.	D	1
10.	D	1
11.	B	1
12.	C	1
13.	D	1
14.	B	1
15.	A	1
16.	C	1
17.	B	1
18.	A	1
19.	B	1
20.	D	1

Section 2

Question			Expected response	Max mark
1.	(a)	1	• DNA contains thymine **AND** RNA contains uracil **OR**	2
		2	• DNA contains deoxyribose **AND** RNA contains ribose **OR** • DNA is double stranded **AND** RNA is single stranded *Any 2, 1 mark each*	
	(b)		• Adds complementary RNA nucleotides to the primary RNA transcript **OR** unwinds and unzips the double helix of DNA	1
	(c)		• Introns are removed from the primary RNA transcript = 1 • Exons are joined together to make mature mRNA = 1	2
	(d)		• Carries a complementary copy of the DNA codes from the nucleus to the ribosome	1

Question			Expected response	Max mark
2.	(a)		• CUCG	1
	(b)		• Label at longer end of RNA chain	1
	(c)		• Anticodon = 1 • Ensure that the correct/specific amino acid is added to the polypeptide = 1	2
3.	(a)	(i)	• Allopatric	1
		(ii)	• Different sequences arise by mutation on either side of the barrier =1 • Barrier prevents gene flow = 1	2
	(b)	(i)	• Hybrid zones **OR** • Zones of hybridisation	1
		(ii)	• Different species could not interbreed to produce fertile young	1
4.	(a)		• Correct scales **AND** labels = 1 • Correct plots **AND** straight line connections = 1	2
	(b)	(i)	• Time taken for decolourisation of resazurin	1
		(ii)	• Concentration/volume of yeast suspension • Concentration/volume of glucose • Concentration/volume of resazurin *Any 2 = 1*	1
	(c)		• 5cm³ glucose • 5cm³ resazurin • (5cm³ distilled water)	1
	(d)		• To allow all solutions to reach the temperature of the water bath so that the reaction would occur at the appropriate temperature	1
	(e)		• 150s	1
	(f)		• Hydrogen ions flow through ATP synthase	1
5.	(a)		• T fish • R complete double • Amphibian/reptile incomplete double *All 6 correct = 2, 4 or 5 = 1*	2
	(b)		• X into heart • Y out of heart • Z out of heart *All 3 = 2, 2 = 1*	2
	(c)		• Oxygenated and deoxygenated blood is kept separate = 1 • Allows more oxygen to be delivered to cells = 1	2
	(d)		• Habitat – deep water; high mountains; deep in mud; others = 1 • Adaptation – increase in red blood cell count; increased haemoglobin content; others = 1	2

Question			Expected response	Max mark
6.	(a)	(i)	• 2400%	1
		(ii)	• Increases in raffinose occur as winter approaches preventing damage to needles by low temperature	1
	(b)		• Migration	1
	(c)		• Organisms that live in extreme conditions that most other life forms are unable to tolerate	1
7.	(a)		• To sterilise the fermenter/prevents growth of other microorganisms	1
	(b)		• Mix the contents/prevents settling of contents **OR** • Ensure even distribution of heat/air	1
	(c)		• Provides oxygen for respiration • Provides respiratory substrate • Provides optimum conditions for enzymes *All 3 = 2, 1 or 2 = 1*	2
8.	(a)	(i)	• Ligase	1
		(ii)	• To allow survival/identification/selection of cells which have taken up the modified plasmid	1
		(iii)	• Prokaryote; horizontally; eukaryote *All 3 = 2, 2 or 1 = 1*	2
	(b)		• To avoid polypeptides being folded incorrectly **OR** • Avoid polypeptides lacking post-translational modifications	1
9.	(a)	(i)	• First nymph stage	1
		(ii)	• 2.5 days	1
	(b)	(i)	• 1 As temperature increases average length of egg laying period decreases = 1 • 2 As temperature increases average number of eggs laid decreases = 1	2
		(ii)	• 5 eggs	1
	(c)	(i)	• T • F • F • T *All 4 = 2, 3 or 2 = 1*	2
		(ii)	• 11.6 days	1
10.	A		• 1 Genetic diversity • 2 Genetic diversity is number and frequency of alleles in a population • 3 Species diversity • 4 Species diversity is species richness • 5 And the relative abundance of each species • 6 Ecosystem diversity • 7 Ecosystem diversity is the number of different/distinct ecosystems *Any 5, 1 mark each*	5

Question			Expected response	Max mark
	B		• 1 extinction of many species at one time • 2 indicated by fossil evidence • 3 there have been several/5 events in the past • 4 biodiversity recovers slowly following mass extinction • 5 due to speciation/adaptive radiation of survivors • 6 extinction rates are difficult to measure • 7 current rates suggest a current mass extinction in progress *Any 5, 1 mark each*	5
11.	(a)		• G3P → • Aspartic acid → • Glycine → • Alanine	1
	(b)		• 0.8	1
	(c)	(i)	• In light G3P decreases as RuBP increases	1
		(ii)	• Hydrogen and ATP come from the light dependent stage = 1 • 3 glyceraldehyde phosphate = 1	2
12.	(a)		• Chemical = 1 • Use of natural predator/disease/pathogen of pest = 1 • Crop rotation/deep ploughing = 1	3
	(b)		• Insecticide might damage non-target insects **OR** • Pesticide might accumulate/magnify in the environment	1
13.	(a)	(i)	• Mutualism	1
		(ii)	• Parasitism = 1 • Parasite benefits in energy/nutrients **AND** host is harmed by loss of these resources = 1	2
	(b)		• Share genes with queen's offspring = 1 • Assisting in raising offspring helps ensure the survival of shared genes = 1	2

Question			Expected response	Max mark
14.	A	(i)	• 1 stems cells are unspecialised/undifferentiated cells (in animals) • 2 stem cells (continue to) divide • 3 to form new stem cells • 4 differentiate/develop into specialised cells/different cells for different functions **OR** • differentiated cells cannot return to an undifferentiated state/change into other cell types • 5 there are adult and embryonic stem cells • 6 Embryonic stem cells can differentiate into almost all cell types • 7 Adult (tissue) stem cells can differentiate to form cells characteristic of the tissue in which they originated/give rise to more limited cell types • 8 example of stem cell source *Any 6, 1 mark each*	6
		(ii)	• 9 can be used to replace damaged or diseased cells/organs/tissues • 10 e.g. diabetes, Parkinson's disease, leukaemia, skin graft, corneal graft, bone marrow transplant • 11 stem cell treatment allows moral requirement to alleviate suffering • 12 embryonic stems cells can only be sourced by destruction of embryos *Any 3, 1 mark each*	3
	B	(i)	• 1 genomics involves studying of gene sequences • 2 gene sequences are used to show evolutionary relatedness • 3 evolutionary relatedness is the basis of phylogenetic trees • 4 to add timescales to phylogenetic trees fossils are needed • 5 molecular clocks are based on sequence differences of a particular protein • 6 differences in sequences related to a protein in different species are graphed on one axis • 7 the other axis shows the timescale of divergence based on relative sequence differences *Any 6, 1 mark each*	6

Question			Expected response	Max mark
		(ii)	• 8 personal genomics involves analysing the genome of an individual • 9 disease risk has a genetic component • 10 genomic differences influence the effectiveness of treatments/drugs • 11 so different treatments can be designed for an individual • 12 example of a difficulty with personalised medicine **EITHER** – the understanding of the complex nature of disease **OR** – distinguishing between neutral and harmful mutations *Any 3, 1 mark each*	3

HIGHER FOR CfE BIOLOGY MODEL PAPER 2

Section 1

Question	Response	Mark
1.	C	1
2.	D	1
3.	B	1
4.	D	1
5.	A	1
6.	C	1
7.	B	1
8.	D	1
9.	A	1
10.	C	1
11.	C	1
12.	B	1
13.	A	1
14.	D	1
15.	B	1
16.	A	1
17.	B	1
18.	D	1
19.	C	1
20.	B	1

Section 2

Question		Expected response	Max mark
1.	(a)	• The DNA is heated to separate the strands/heating to 90°C denatures the DNA and separates the strands	1
	(b)	• Primers are needed to allow replication/sequencing to start **OR** • Primers keep the strands separate **OR** • Primers act as a starting point for the enzyme/DNA polymerase	1
	(c)	• DNA polymerase **OR** • Taq polymerase	1
	(d)	• 16	1
2.	(a)	• Exons	1
	(b)	• Primary mRNA transcript	1
	(c)	• Different proteins can be expressed from one gene	1
	(d)	• Adding phosphate **OR** • Carbohydrate groups to the protein	1
3.	(a)	• Unspecialised cells = 1 • Able to divide **OR** • Able to differentiate into other cell types = 1	2

Question			Expected response	Max mark
	(b)	(i)	• 250%	1
		(ii)	• 475	1
		(iii)	• 7 months	1
	(c)		• To provide information on cell processes/cell growth/ differentiation/gene regulation **OR** • Used as model cells to study how diseases develop **OR** • Used as model cells for drug testing	1
4.	(a)		• Sequence data/fossil evidence	1
	(b)		• 73 +/-1	1
	(c)		• 20	1
	(d)		• 5	1
5.	A		• 1 hybrid zones form where the ranges of two closely related species overlap • 2 within hybrid zones members of the two species interbreed • 3 the hybrid offspring are less fit/ sterile • 4 and are eliminated by natural selection • 5 members of each species re-colonise the hybrid zone • 6 and undergo further hybridisation to repopulate the zone *Any 4*	4
	B		• 1 change in frequency of DNA sequences • 2 random process • 3 results of neutral mutations • 4 and founder effects • 5 example of founder effect – change colonisation by sub-populations • 6 particularly affects small populations *Any 4*	4
6.	(a)		• 140	1
	(h)		• At 0 minutes/start the rate of urine production was 2 cm³/minute • Increased to 20 cm³/minute at 90 minutes • Then decreased to 2 cm³/minute at 140 minutes • Then remained constant (to 180 minutes) *All 4 = 2, 3 correct = 1, Units required only once* *All figures correct but no units =1*	2
	(c)		• 60-80	1
	(d)		• 14	1
	(e)		• Analysis of an individuals genome/ knowledge of the genetic component of risk of disease	1
7.	(a)		• Volume of extract/solution/sample • pH of solutions • Time left in colorimeter/out of bath *Any 2, 1 mark each*	2

Question		Expected response	Max mark
	(b)	• Some reaction occurred/Enzyme started working/Browning occurred/ Pigment produced **AND** • immediately before lead added/ before reading taken/as soon as cut/ while being cut up **OR** • Apple tissue was already damaged	1
	(c)	• Appropriate enclosed scales with zeros and labels exactly from table **AND** • Shared zero must be exactly at origin = 1 • Points plotted accurately and joined with straight lines **AND** • Line labelled from table = 1	2
	(d)	• As concentration/it increases, the activity (of the enzyme) decreases/ effect of enzyme decreases/the enzyme is inhibited more **OR** • As concentration/it increases, the rate of browning decreases **OR** • Converse	1
	(e)	• Repeat experiment exactly the same with test tube containing sample of extract + water	1
	(f)	• Repeat experiment and obtain an average	1
8.	(a)	• NAD **AND** FAD *Both for 1*	1
	(b)	• Provide the energy/are used to pump hydrogen ions across the membrane	1
	(c)	• ATP synthase	1
	(d)	• The final electron acceptor and combines with hydrogen ions and electrons to form water	1
9.	(a)	• Blood vessel – constricts so reduced flow of blood to skin surface and less heat loss by radiation **OR** • Sweat gland – reduced sweat production and less heat loss by evaporation **OR** • Hair erector muscle – contracts raising hair and to trap a layer of insulating air	1
	(b)	• Optimal temperature for enzyme controlled reaction rates = 1 **AND** • Optimal diffusion rates = 1	2
	(c)	• Conformers	1

Question			Expected response	Max mark
10.	(a)		• Q and S	1
	(b)		• It provides a way/method of introducing the foreign DNA/gene	1
	(c)		• Ligase	1
	(d)		• Reproduce rapidly producing identical GM cells/clones/cells containing the required gene	1
	(e)		• Polypeptides can be folded incorrectly **OR** • Polypeptides may lack post translational modifications = 1 • Yeast = 1	2
11.	(a)		• To eliminate any effects of contaminating microorganisms	1
	(b)		• Temperature/oxygen/pH *Any 2 for 1*	1
	(c)		• 0.5	1
	(d)		• Antibiotics = 1 • Inhibit the growth of other species of bacteria and so reduce competition for available resources = 1	2
12.	(a)	(i)	• Naturalised	1
		(ii)	• May be free of predators/parasites/pathogens/competitors/ability to outcompete native species	1
	(b)	(i)	• Insecticide only - the population of fire ants decreased by 90%/fell to 10% 1 year after treatment then started to increase = 1 • Integrated approach - the population of fire ants decreased by 95%/fell to 5% 1 year after treatment and then remained constant = 1	2
		(ii)	• Some fire ants/those with resistance to the insecticide survive = 1 • They reproduce and pass on their resistance/favourable/beneficial gene to their offspring =1	2
		(iii)	• Introduced predator may become invasive/effect other native organisms/disrupt food web/reduce biodiversity	1
		(iv)	• The phorid fly larvae/parasite benefits in terms of energy/nutrients where as the host/fire ant is harmed	1
13.	(a)		• To eliminate genetic differences having an effect on the results **OR** • So that temperature was the only factor affecting results	1
	(b)	(i)	• As the temperature increased from 0°C to 20°C the mean growth rate increased = 1 • As temperature increases above 20°C the mean growth rate decreases = 1	2
		(ii)	• 5:6	1

Question			Expected response	Max mark
		(iii)	• All other variables should be kept constant or named examples/mass of food/type of food/quality of food/time of feeding/level of activity	1
		(iv)	• Increased heat loss at low/0°C and so more energy from food being used to maintain body temperature	1
		(v)	• Stereotype/misdirected behaviour/failure in sexual behaviour/failure in parental behaviour/altered levels of activity	1
14.	A	(i)	• 1 measurable components include genetic diversity, species diversity, ecosystem diversity *(any 2)* • 2 a third component (from the above) • 3 genetic diversity is the number **AND** frequency of all alleles/genetic sequences in a population • 4 species diversity is the species richness **AND** relative abundance • 5 species richness is the number of species • 6 relative abundance is the proportion of each species • 7 ecosystem diversity is variety/mosaic of habitats/number of different ecosystems *Max 4 (from 7)*	4
		(ii)	• 8 overexploitation • 9 habitat loss reduces biodiversity **OR** • example • 10 habitat fragmentation reduces biodiversity **OR** • example • 11 invasive/introduced species eliminate/outcompete native species and reduce biodiversity • 12 invasive species may be free of predators/pathogens which naturally limit their populations **OR** example • 13 invasive species may outcompete/prey on native species • 14 climate change is threat to biodiversity *Max 5 (from 7)*	5
	B	(i)	• 1 food security is ability of human population to access sufficient quantity/amount of food • 2 and sufficient quality of food • 3 increasing population increases the demand for food production • 4 demand that production is sustainable • 5 and does not degrade natural resources on which agriculture depends *Max 3 (from 5)*	3

Question			Expected response	Max mark
	(ii)		• 6 food production depends on photosynthesis • 7 food production can be increased by planting increased area of crop • 8 factors which limit photosynthesis/control plant growth • 9 include light, temperature, CO_2 availability • 10 high-yielding cultivars/GM crops • 11 protection of crops from disease/pests/competition • 12 increased irrigation/fertilisers • 13 livestock produce less food per unit area than crops • 14 due to energy loss at each trophic level • 15 livestock production may be possible in wild habitats not suitable for cultivation *Max 6 (from 10)*	6

HIGHER FOR CfE BIOLOGY MODEL PAPER 3

Section 1

Question	Response	Mark
1.	C	1
2.	B	1
3.	A	1
4.	A	1
5.	D	1
6.	B	1
7.	C	1
8.	A	1
9.	D	1
10.	A	1
11.	B	1
12.	C	1
13.	B	1
14.	D	1
15.	B	1
16.	D	1
17.	D	1
18.	C	1
19.	C	1
20.	D	1

Section 2

Question			Expected response	Max mark
1.	(a)		• Transcription	1
	(b)		• A sequence of 3 bases codes for a specific amino acid	1
	(c)	(i)	• One strand runs form 3' to 5' and its complementary strand runs from 5' to 3'	1
		(ii)	• Removes introns and joins exons together	1
		(iii)	• An intron may be left in the mature mRNA molecule	1
2.	(a)		• Peptide	1
	(b)		• Hydrogen bond/disulphide bridge	1
	(c)		• Active site shape is complementary/fits into specific substrate	1
	(d)		• Active site changes shape to match and bind substrate molecule	1
3.	(a)		• 1 and 2	1
	(b)		• Vertebral column, hinged jaws, 4 walking legs	1
	(c)		• 4	1
	(d)	1	• Fossil record	1
		2	• Sequence data	1
4.	(a)		• That protein synthesis occurs at the ribosomes	1

Question			Expected response	Max mark
	(b)		• From 0 to 4 minutes the mass of amino acids increases = 1 • Between 4 and 5 minutes the mass remains constant = 1	2
	(c)		• 120	1
5.	(a)	(i)	• A, C	1
		(ii)	• B, H	1
		(iii)	• F, G	1
		(iv)	• F, G	1
	(b)		• Cristae/inner membranes of mitochondria	1
6.	A		• 1 phospholipid bilayer • 2 proteins embedded • 3 membranes form compartments/localise the metabolic activity of the cell • 4 provide more favourable conditions for reactions to take place • 5 membrane folds in organelles/mitochondria/chloroplasts provide a large surface area for metabolic reactions to take place • 6 the large surface to volume ratio of small compartments allow high concentrations of reactants to occur/higher reaction rates possible *Any 4, 1 each*	4
	B		• 1 competitive inhibitors block active sites • 2 competitive inhibitors slow down reactions • 3 competitive inhibitors are affected by substrate concentration • 4 non-competitive inhibitors bind to allosteric site/site other than active site • 5 non-competitive inhibitors slow down reactions irreversibly • 6 non-competitive inhibitors are not affected by substrate concentration *Any 4, 1 each*	4
7.	(a)	(i)	• Beginning of August until end of August increases from 4.0 – 4.5 grams • End August until end November decreases from 4.5 – 1.5 grams • From end November until end January remains at 1.5 grams *All 3 = 2, 2 = 1*	2
		(ii)	• 60%	1
		(iii)	• More food/nectar available	1
	(b)	(i)	• Saves energy for flying	1
		(ii)	• 0.5	1
	(c)		• 45.0	1
	(d)		• Ringing/banding/marking **AND** recovery **OR** • Radio tracking/tagging	1
8.	(a)	(i)	Y = 1, W = 1	2

Question			Expected response	Max mark
		(ii)	• gives organisms a competitive advantage = 1 • required to metabolise available substrate = 1	2
	(b)		• To allow the huge numbers involved to be plotted on a small section of graph paper	1
	(c)		• Temperature, oxygen level, pH *Any 2, 1 each*	2
9.	(a)	(i)	• Light intensity	1
		(ii)	• Temperature • Concentration of Chlorella • Concentration of CO_2 • pH *Any 2*	1
	(b)	(i)	• To allow time for the Chlorella to acclimatise/to be carrying out photosynthesis under the experimental conditions/time for photosynthesis to take place/time for intermediates to be produced	1
		(ii)	• Allows easier sampling/extraction (of intermediates)	1
	(c)		• Repeat the experiment with a further 500cm³ of Chlorella and obtain an average	1
	(d)		• Correct scales **AND** labels = 1 • Correct plots **AND** straight line connections = 1	2
	(e)		• In the light/when illuminated the RuBP concentration is low/8 units and the G3P concentration is high/20 units = 1 • When the light is switched off/in the dark the G3P concentration decreases and the RuBP concentration starts to increase = 1	2
10.	(a)		• Ethogram	1
	(b)		• mouth open/wide and teeth visible = 1 • mouth slightly open/narrow/closed and teeth not visible = 1	2
	(c)	(i)	• Cooperative hunting/defence is more effective in a group	1
		(ii)	• Time to learn complex social behaviours/skills	1
		(iii)	• Reduces aggression/unnecessary conflict/saves energy/formation of alliances	1
11.	(a)		• May be free of predators/parasites/pathogens/competitors which occur in their native habitat	1
	(b)		• May prey on native species • Outcompete native species for resources • May hybridise with native species • Habitat degradation • Increased grazing pressures *Any 2, 1 each*	2

Question			Expected response	Max mark
	(c)		• Method; introduce natural predator/ pathogen/disease = 1 • Drawback; introduced predator may become invasive/effect other native organisms/disrupt food web/reduce biodiversity **OR (must match)** • Disease/(microbial) pathogen may affect native/non-target species	2
12.	(a)		• The average milk yield of the F_1 crossbreed population is greater than that of both of the pure breed populations	1
	(b)	(i)	• The F_2 population/individuals/ generation are genetically variable/ have a wide variety of genotypes and have a reduced/lower milk yield	2
		(ii)	• Livestock produce less food per unit area than plant crops/energy is lost between trophic levels	1
	(c)		• Test cross; used to identify unwanted individuals with heterozygous recessive alleles = 1 • Back cross; required to maintain the new breed = 1	2
13.			• Meaning; a behaviour that harms the donor individual but benefits the recipient = 1 • Explanation; donor/birds benefit in terms of increased chances of survival of shared genes • In the recipients/related birds offspring **OR** • Their own future offspring will similarly benefit by reciprocal behaviour = 1 *Any 2, 1 each*	3
14.	A	(i)	• 1 genetic drift is random increases and decreases in frequency of sequences/alleles/genes • 2 particularly in small populations • 3 results of neutral mutations • 4 (or as results of the) founder effect/principle • 5 founders of new populations have gene sequences which are not representative of the whole population/are in abnormal percentages of the whole population *Any 3, 1 each*	3

Question			Expected response	Max mark
		(ii)	• 1 organisms show (genomic) variation (upon which natural selection acts) • 2 natural/sexual selection is non-random • 3 (causes) increase in frequency of certain genetic sequences/genes/ alleles • 4 these sequences/alleles increase survival **OR** • survival of the fittest/best suited to their environment • 5 (survivors) pass on their favourable/ beneficial gene sequence/alleles/ characteristics to offspring/next generation • 6 deleterious/damaging sequences/ alleles/genes/characteristics are reduced in frequency/removed from the population • 7 stabilising selection (and explanation) directional selection (and explanation) disruptive selection (and explanation) *Any 1, must match* • 8 sexual selection • 9 over many generations/a long time • individual with favourable/beneficial characteristics/genes/alleles survive to breed/reproduce *Any 6, 1 each*	6
	B	(i)	• in prokaryotic cells; • 1 prokaryotic cells do not have a distinct nucleus • 2 DNA is packaged in circular chromosomes • 3 and in plasmids in eukaryotic cells; • 4 eukaryotic cells have a nucleus • 5 DNA is packaged in linear chromosomes • 6 and in circular chromosomes in chloroplasts and mitochondria *Any 4, 1 each*	4
		(ii)	• 1 the DNA is heated to 90°C/95°C • 2 this denatures the DNA/which separates the strands • 3 temperature is cooled/lowered to 55°C/60°C • 4 allows primers to bind to the target sequences • 5 the temperature is then raised to over 70°C • 6 when heat tolerant DNA polymerase is used to synthesise new strands • 7 from free DNA nucleotides. • 8 repeated cycles allow millions of copies of the target sequence to be produced. *Any 5, 1 each*	5

HIGHER FOR CfE BIOLOGY 2015

Section 1

Question	Answer	Mark
1.	D	1
2.	B	1
3.	D	1
4.	A	1
5.	D	1
6.	B	1
7.	C	1
8.	C	1
9.	B	1
10.	C	1
11.	D	1
12.	A	1
13.	A	1
14.	D	1
15.	B	1
16.	C	1
17.	A	1
18.	B	1
19.	C	1
20.	A	1

Section 2

Question			Expected answer(s)	Max mark
1.	(a)	(i)	• ADP + Pi/phosphate/inorganic phosphate *Both required for 1 mark*	1
		(ii)	• NAD	1
		(iii)	• It is a net/overall energy gain (following an energy investment at an earlier stage) OR • More ATP/energy is produced/released than is used/invested (earlier/in stage1)	1
	(b)	(i)	• Increases the surface area for (action of) bacteria/Lactobacillus OR • Bursts cells to release more substrate/cell contents for bacterial action	1

Question			Expected answer(s)	Max mark
		(ii)	• Acidic conditions/low pH/anaerobic conditions • Inhibits/kills other bacteria OR • pH/oxygen levels optimum for Lactobacillus but not for other bacteria	1
2.	(a)		• Same/complementary sticky ends OR • Complementary/matching base sequence/bases	1
	(b)		• (DNA) ligase	1
	(c)		• In the presence of antibiotic only these/modified bacteria grow/survive OR • Converse	1
	(d)		• Origin of replication	1
	(e)	(i)	• Lack of post-translational modifications OR • Protein/polypeptide is folded incorrectly	1
		(ii)	• Put modified plasmid into yeast OR • Chemically modify protein	1
3.	(a)	(i)	• Lactose concentration	1
		(ii)	• Temperature/concentration of yeast/pH	1
	(b)		• As lactose increases (from 4%) to 16%, ethanol concentration increases 1 • From 16% (to 20%) ethanol remains constant/levels off 1 • As lactose (concentration) increases, ethanol (concentration) increases then levels off = 1 mark if neither above points included	2
	(c)		• 37.5	1
	(d)		• Aerobic respiration does not produce ethanol OR • Aerobic respiration produces water not ethanol OR • No/less anaerobic respiration so less ethanol produced	1
4.	(a)		• 112	1
	(b)		• 08:00—12:00 1 • Time of lowest metabolic rate 1	2

	Question		Expected answer(s)	Max mark
	(c)		• Energy saved/conserved OR • Uses less energy OR • Energy not wasted	1
	(d)		• Dormancy OR • Hibernation OR • Aestivation OR • A correct description of one of these terms	1
5.	A		• 1 Amphibian heart has 2 atria and 1 ventricle1 1 • 2 Bird heart has 2 atria and 2 ventricles 1 • 3 Birds have a higher metabolic rate (or converse) 1 • 4 No mixing of oxygenated and deoxygenated blood in bird heart (or converse) 1 • 5 More efficient oxygen delivery to bird cells/muscles/tissues (or converse) 1	4
	B		• 1 A competitive inhibitor binds to/blocks the active site 1 • 2 Competitive inhibition is reversed/reduced by increasing substrate concentration 1 • 3 Non-competitive inhibition is where a molecule binds to the enzyme not on the active site 1 • 4 Non-competitive inhibitor changes (the shape of) the active site 1 • 5 Non-competitive inhibition is irreversible/not affected by substrate concentration 1	4
6.	(a)	(i)	• Enzyme X: RuBisCO 1 • Substance Y: G3P/ glyccraldchydc 3 phosphatc 1	2
		(ii)	• Glucose: for respiration/ ATP (production)/cellulose formation/starch formation/ other biosynthetic pathways/ processes 1 • RuBP: for continuation of the cycle/to allow cycle to occur/ repeat 1	2
	(b)	(i)	• ATP: increases 1 • NADPH: increases 1	2

	Question		Expected answer(s)	Max mark
		(ii)	• Procedure: randomisation of plots/treatments 1 • Explanation: reduces/ eliminates bias 1 OR • Procedure: replication/number of replicates • Explanation: to take account of variability/reduce the effect of atypical results OR • Procedure: selection of treatments/inclusion of both GM and non GM crops • Explanation: to make/ensure a (fair) comparison	2
7.	(a)		• Parasite/Schistosoma gets energy/gets nutrients AND • Host/human is harmed (by loss of resources)	1
	(b)		• Secondary host: Fresh water snails 1 • Benefit: allows development into free swimming parasite 1 OR • Allows them/immature parasites to complete life cycle	2
	(c)		• Prevent urine/faeces/eggs from entering (fresh) water OR • Stop people entering the affected water OR • Control the population of fresh water snails OR • Medication given to kill the eggs/ mature parasite/parasite in humans	1
8.	(a)		• Biological (control)	1
	(b)		• Harlequin ladybird has spread (rapidly) AND • Native populations/ladybirds are decreasing (and may be eliminated)	1
	(c)		• May have alternative prey/ food source/niche/resources OR • Not competing with the Harlequin ladybird OR • Less competition with other native species OR • Not preyed upon by Harlequin ladybirds	1
	(d)		• Free from its usual predators/ parasites/pathogens/disease/ competitors	1

Question			Expected answer(s)	Max mark
9.	(a)	(i)	• 1. From 0 – 60 kg per hectare increase from 3 – 8·4 tonnes per hectare • 2. Remains at 8.4 between 60 and 80 • 3. Between 80 and 100 decreases from 8.4 – 7.9/7.8	2
		(ii)	• 208 – 216	1
		(iii)	• 840	1
	(b)	(i)	• Use of 10 cattle in each group	1
		(ii)	• 0·6	1
		(iii)	• 20 • Increasing the phosphate/ fertiliser level increases the growth rate	1 1
	(c)		• Energy lost at each level/ stage of a food chain **OR** • Energy lost between trophic level(s)	1
10.	(a)		• Horizontal 1 • Rapid 1	2
	(b)		• Resistant survive (or converse) 1 **AND** • Transfer/pass on the (antibiotic) resistance gene/ allele to next generation 1 **OR** • Transfer/pass on resistance gene vertically/horizontally	2
	(c)	(i)	• Growth/bacteria/MRSA/ colonies would be present on all plates	1
		(ii)	• Vitamins/fatty acids/beef extract	1
11.	(a)		• 0.24	1
	(b)		• 32	1
	(c)		• Inclusive scale and axes labels copied exactly from table headings 1 • Points plotted and joined with a ruler 1	2
	(d)		• Only donor 2 is suitable **OR** • Donor 2 is most suitable	1
	(e)	(i)	• TACTGTTTAGC	1
		(ii)	• Separates DNA strands/breaks H bonds between strands/ denatures DNA 1 • Any temperature from 50 – 65°C 1	2

Question			Expected answer(s)	Max mark
12.	(a)		• Increase in stroke volume/ volume of blood pumped out of heart per heartbeat 1 • No effect on heart rate 1	2
	(b)		• 1190	1
	(c)	(i)	• Embryonic stem cells differentiate/develop into all types of cell **AND** • Adult/tissue stem cells differentiate/develop into narrower range of cell types **OR** • Adult stem cells are more differentiated/specialised than embryonic stem cells	1
		(ii)	• They express/switch on the genes characteristic of that type of cell **OR** • Certain genes/some genes are expressed/switched on and other genes are switched off	1
	(d)		• Provide information on gene regulation/cell growth/ cell differentiation/cell division/cell ageing/disease development	1
13.	A	(i)	• 1 RNA polymerase unzips/ unwinds DNA or separates DNA into two strands 1 • 2 Hydrogen bonds between strands/base pairs break 1 • 3 RNA polymerase aligns/ brings in/joins/attaches RNA nucleotides with their complementary nucleotides/ bases on DNA (template) or A to U and T to A and C to G in diagram 1 • 4 a primary transcript is produced 1 • 5 exons are coding and introns are non-coding (regions of the primary transcript) 1 • 6 introns/non-coding regions are removed OR exons/coding regions are retained 1 • 7 exons are spliced/joined together to form mature mRNA)/transcript 1	5

Question			Expected answer(s)	Max mark
		(ii)	• a. tRNA has an anticodon and an amino acid attachment site 1 • b. tRNA binds/joins to/carries/ collects specific/correct amino acid 1 • c. tRNA carries (specific) amino acid to ribosomes 1 • d. anticodons are complementary/pair with codons on mRNA 1 • e. there are start and stop codons 1 • f. peptide bonds form between amino acids 1 **OR** a polypeptide forms	4
B	(i)		• 1 (single gene) mutations are random changes in DNA sequences/genes/ alleles/the genome 1 • 2 single gene mutation name AND description Substitution – base/base pair/ nucleotide is replaced/ substituted by another Insertion – base/base pair/ nucleotide is added/inserted Deletion – base/base pair/ nucleotide is removed/ deleted 1 • 3 another single gene mutation name AND description 1 • 4 If 2 or 3 not awarded – all 3 mutation names 1 • 5 Insertion/deletion results in a frameshift mutation/expansion of a nucleic acid sequence 1 • 6 (single gene) mutations are important in evolution 1 • 7 splice site mutations can alter the mature mRNA OR result in exon removal OR result in introns remaining present 1	4

Question			Expected answer(s)	Max mark
		(ii)	• a. chromosome mutation can involve changes to chromosome number/structure 1 • b. chromosome mutation name AND description; Translocation: genes/sections of chromosome from one chromosome become attached to another chromosome Deletion: genes/sections of chromosome deleted from chromosome Inversion: genes/sections of chromosome/rotate through 1800/flipped Duplication: genes/sections of chromosome/pieces of chromosome are duplicated/ repeated 1 • c. another chromosome mutation name AND description 1 • d. If b or c not awarded - all 4 names but no descriptions 1 • e. polyploidy results from errors during the separation of chromosomes/non-disjunction/ spindle failure during cell division/ meiosis/mitosis/ gamete formation 1 • f. polyploidy is the possession of complete extra sets/ double/triple the number of chromosomes OR a whole genome duplication OR 2n becomes 3n 1 • g. polyploidy is important in evolution of food crops OR duplication provides material upon which natural selection can work/is important in evolution 1 • h. polyploid crops/plants/ named example show desirable features/ higher yields/other appropriate examples of desirable features 1	5

Acknowledgements

Hodder Gibson would like to thank SQA for use of any past exam questions that may have been used in model papers, whether amended or in original form.